A GENERATION ON THE MARCH

THE UNION ARMY AT GETTYSBURG

by
Edmund J. Raus, Jr.

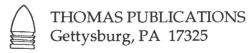

THOMAS PUBLICATIONS
Gettysburg, PA 17325

Dedicated to my parents
Edmund J. and Marion Curtis Raus

A GENERATION ON THE MARCH

THE UNION ARMY
AT GETTYSBURG

CONTENTS

ACKNOWLEDGEMENTS

At the top of my list of those to acknowledge for assistance in the writing of this book are Roger Hunt of Rockville, Maryland and Mike Musick of the National Archives in Washington, D.C. Mr. Hunt has no equal in tracking down biographical information on Union Civil War officers. Mr. Musick's name is familiar to many students of the war who have benefited from his knowledge of the vast Civil War holdings of the National Archives.

I wish to also thank historians Robert K. Krick of Fredericksburg and Spotsylvania National Military Park and Kathy Georg Harrison of Gettysburg National Military Park. Both looked at the typescript and offered helpful suggestions. A nod of appreciation is also due Russ Pritchard, MOLLUS War Library and Museum; Michael J. Winey, U.S. Army Military History Institute; and Greg Coco, Robert H. Prosperi, and Scott Hartwig. The maps for this book were drawn by Woody Christ, a skilled cartographer and Gettysburg licensed battlefield guide.

We are people to whom the past is forever speaking. We listen to it because we cannot help ourselves, for the past speaks to us with many voices. Far out of that dark nowhere which is the time before we were born, men who were flesh of our flesh and bone of our bone went through fire and storm to break a path to the future....

Bruce Catton, *America Goes to War*

INTRODUCTION

While working behind the information desk at Gettysburg National Military Park (1974-1977), I was surprised at the number of park visitors seeking information on a hometown regiment or one in which a relative fought. In the summer of 1863 close to 200,000 Americans, representing a cross section of that generation, converged on Gettysburg. Today, surviving battle accounts, service records and unit histories leave a trail for their descendants to follow. Arriving on the battlefield, their search is aided by the many unit monuments and markers erected primarily by veterans' groups to mark battle positions. By locating these guideposts, visitors can literally walk in the footsteps of their ancestors.

This book is offered as a guide to facilitate the search on the Union side of the roster. Each of the 345 Union units at Gettysburg has an entry here, arranged by state. The brigade, division and corps association of each unit is noted as well as any common nickname. An attempt is made to provide the places of recruitment and the place of organization as a regiment or battery. Biographical information is provided only for the 469 unit commanders at Gettysburg, July 1-3. Numbers and losses for each regiment are given, and, if a monument exists, the text identifies what description of the fighting it provides. An index lists the counties where the units were raised and the names of the commanders. Finally, maps are provided to assist in locating the monuments and markers of the units that fought on the battlefield. It is my hope that there will be something here for both the Civil War buff and the casual visitor who might be interested in a particular regiment or in the individual who commanded it.

Further explanation is needed for many of the above categories. The date of organization refers to the date the unit was mustered into Federal service. State service or any period of three-months service is ignored. Since the date of muster does not always coincide with the place where the unit was organized, these facts are separated by a period. Along with the year of organization for prewar Regular units, I include the location of the unit at the time Fort Sumter was fired upon, April 12, 1861.

At the time of the Battle of Gettysburg, many United States Regular artillery batteries were still officially organized as companies in regiments. For the sake of consistency and to avoid confusion, the term "battery" is used throughout instead of "company."

In trying to determine where the units were recruited, I decided to provide only the 367 counties (in a few cases cities) that I believed were substantially represented in units present on the field at Gettysburg. Most often this means at least a company strength at enlistment for infantry and cavalry units and 25 or more men for artillery batteries. The Regular outfits posed a particularly difficult problem. Muster rolls indicate where a man enlisted. Since many men were sent or traveled to centralized depots to enlist, the locations (usually larger cities) do not necessarily represent their residences. When the Regular Army expanded in 1861, the competition for men drove recruiting officers into areas beyond the big cities to fill the ranks of the new units. Still, I cannot always be sure the men were from the particular counties. For example, when looking over several unit muster rolls, I was confused by the great number of men enlisting in the small towns along the St. Lawrence River in northern New York. I finally concluded that many must have been Canadians or others coming across the border to join the Union cause.

Every unit commander at Gettysburg has a life sketch. For each I tried to find date and place of birth; any indication of higher education, militia service or Mexican War service; place of residence and occupation at the start of the Civil War; service record; major accomplishments after the war; and date and place of death. Burials in National Cemeteries also are noted. Service information in the Gettysburg unit includes initial rank and earliest authorized date of rank, date of any wounds, and date and rank at muster out. All officers are shown being mustered into specific companies, batteries or regiments, although sometimes the units did not receive their official designations until later. For example, Capt. Benjamin L. Higgins was mustered in with his company, the Steuben Rangers, on November 12, 1861. On November 23, 1861, the company became Company A of the 86th New York Infantry. Higgins' entry in this study has him Capt., Co. A, 86th N Y., Nov. 12, 1861. No higher rank given at muster out indicates same rank as at Gettysburg.

Numbers and losses for units in a battle are always slippery calculations. Different figures are arrived at according to researchers' sources and perspective. How, for instance, could you accurately determine the number of men that straggled on the long march to Gettysburg and how many returned to the ranks during the four day stay of the army at Gettysburg. Where possible, I use the figure "present for duty" according to the official accounting of men conducted on June 30, 1863. Losses of the Gettysburg units (July 1 - July 4) were taken from *The War of the Rebellion: A Compilation of the Official Records of the Union and Confederate Armies* (Washington, D.C., 1880-1901), Series 1, Vol. 27, part 1, pg. 173.

Please address any additions or comments to the author in care of the publisher.

ABBREVIATIONS

Adjt. ....Adjutant
attnd. ....attended college or university,
but did not necessarily graduate
Art. ....Artillery
b. ....born
brig. ....brigade
Brig. Gen. ....Brigadier General
Capt. ....Captain
Cav. ....Cavalry
co. ....company
Col. ....Colonel
Corpl. ....Corporal
d. ....death
div. ....division
grad. ....graduate
Inf. ....Infantry
indpt. ....independent
k. ....killed
Lt. ....Lieutenant
2 Lt. ....Second Lieutenant
Lt. Col. ....Lieutenant Colonel
m. ....missing
Maj. ....Major
M.I. ....Mustered in to service
M.O. ....Mustered out of service
mort. wd. ....mortally wounded
mos. ....months
pdr. ....pounder as in 12-pounder Napoleon gun
Pvt. ....Private
retd. ....Retired
Sergt. ....Sergeant
U.S.S.S. ....United States Sharpshooters
vol. ....volunteer organization
wd. ....wounded

3

Organization of the Army of the Potomac, Major General George G. Meade, United States Army, Commanding, at the Battle of Gettysburg, July 1-3, 1863.

General Headquarters

Brig. Gen. Marsena R. Patrick
1st Pennsylvania Cavalry
2nd Pennsylvania Cavalry
6th Pennsylvania Cavalry, Companies E and I

Guards and Orderlies

Oneida Independent Company, New York Cavalry

First Army Corps

Maj. Gen. John F. Reynolds
Maj. Gen. Abner Doubleday
Maj. Gen. John Newton

General Headquarters

1st Maine Cavalry, Company L
121st Pennsylvania Infantry, Company B

First Division

Brig. Gen. James S. Wadsworth

First Brigade	Second Brigade
Brig. Gen. Solomon Meredith	Brig. Gen. Lysander Cutler
19th Indiana	7th Indiana
24th Michigan	76th New York
2nd Wisconsin	84th New York
6th Wisconsin	95th New York
7th Wisconsin	147th New York
	56th Pennsylvania

Second Division

Brig. Gen. John C. Robinson

First Brigade	Second Brigade
Brig. Gen. Gabriel R. Paul	Brig. Gen. Henry Baxter
16th Maine	12th Massachustts
13th Massachusetts	83rd New York
94th New York	97th New York
104th New York	11th Pennsylvania
107th Pennsylvania	88th Pennsylvania
	90th Pennsylvania

Third Division

Brig. Gen. Thomas A. Rowley
Maj. Gen. Abner Doubleday

Provost Guard

149th Pennsylvania Infantry, Company D

First Brigade	Second Brigade	Third Brigade
Col. Chapman Biddle	Col. Roy Stone	Brig. Gen. George J. Stannard
Brig. Gen. Thomas A. Rowley		
80th New York	143rd Pennsylvania	13th Vermont
121st Pennsylvania	149th Pennsylvania	14th Vermont
142nd Pennsylvania	150th Pennsylvania	15th Vermont
151st Pennsylvania		16th Vermont

First Corps Artillery

Col. Charles S. Wainwright
5th Maine Artillery (Battery E)
2nd Maine Artillery (Battery B)
1st New York Artillery, Battery L
1st Pennsylvania Artillery, Battery B
4th United States Artillery, Battery B

Second Army Corps

Maj. Gen. Winfield S. Hancock
Brig. Gen. John Gibbon

General Headquarters

6th New York Cavalry, Companies D and K
53rd Pennsylvania Infantry, Companies A, B, and K

First Division

Brig. Gen. John G. Caldwell

Provost Guard

116th Pennsylvania Infantry, Company B

First Brigade	Second Brigade
Col. Edward E. Cross	Col. Patrick Kelly
5th New Hampshire	28th Massachusetts
61st New York	63rd New York
81st Pennsylvania	69th New York
148th Pennsylvania	88th New York
	116th Pennsylvania

Third Brigade	Fourth Brigade
Brig. Gen. Samuel K. Zook	Col. John R. Brooke
52nd New York	27th Connecticut
57th New York	2nd Delaware
66th New York	64th New York
140th Pennsylvania	53rd Pennsylvania
	145th Pennsylvania

Second Division

Brig. Gen. John Gibbon
Brig. Gen. William Harrow

Provost Guard

1st Minnesota Infantry, Company C

Unattached

Massachusetts Sharpshooters, 1st Company

First Brigade	Second Brigade	Third Brigade
Brig. Gen. William Harrow	Brig. Gen. Alexander S. Webb	Col. Norman J. Hall
19th Maine	69th Pennsylvania	19th Massachusetts
15th Massachusetts	71st Pennsylvania	20th Massachusetts
1st Minnesota	72nd Pennsylvania	59th New York
82nd New York	106th Pennsylvania	42nd New York
		7th Michigan

Third Division

Brig. Gen. Alexander Hays

Provost Guard

10th New York Battalion

First Brigade	Second Brigade	Third Brigade
Col. Samuel S. Carroll	Col. Thomas A. Smyth	Col. George L. Willard
14th Indiana	14th Connecticut	39th NewYork
4th Ohio	1st Delaware	111th New York
8th Ohio	12th New Jersey	125th New York
7th West Virginia	108th New York	126th New York

Second Corps Artillery
Capt. John G. Hazard

1st New York Artillery, Battery B
1st Rhode Island Artillery, Battery A
1st Rhode Island Artillery, Battery B
1st United States Artillery, Battery I
4th United States Artillery, Battery A

Third Army Corps
Maj. Gen. Daniel E. Sickles
Maj. Gen. David B. Birney

General Headquarters
6th New York Cavalry, Company A

First Division
Maj. Gen. David B. Birney
Brig. Gen. J. H. Hobart Ward

First Brigade	Second Brigade	Third Brigade
Brig. Gen. Charles K. Graham	Brig. Gen. J. H. Hobart Ward	Col. P. Regis De Trobriand
	Col. Hiram Berdan	
57th Pennsylvania	20th Indiana	17th Maine
68th Pennsylvania	3rd Maine	3rd Michigan
114th Pennsylvania	4th Maine	5th Michigan
63rd Pennsylvania	86th New York	40th New York
105th Pennsylvania	124th New York	110th Pennsylvania
141st Pennsylvania	99th Pennsylvania	
	1st United States S.S.	
	2nd United States S.S.	

Second Division
Brig. Gen. Andrew A. Humphreys

First Brigade	Second Brigade	Third Brigade
Brig. Gen. Joseph B. Carr	Col. William R. Brewster	Col. George C. Burling
1st Massachusetts	70th New York	2nd New Hampshire
11th Massachusetts	71st New York	5th New Jersey
16th Massachusetts	72nd New York	6th New Jersey
12th New Hampshire	73rd New York	7th New Jersey
11th New Jersey	74th New York	8th New Jersey
26th Pennsylvania	120th New York	115th Pennsylvania

7

Third Corps Artillery

Capt. George E. Randolph
1st New Jersey Artillery, Battery B
1st New York Artillery, Battery D
4th New York Artillery, Independent Battery
1st Rhode Island Artillery, Battery E
4th United States Artillery, Battery K

Fifth Army Corps

Maj. Gen. George Sykes

General Headquarters

12th New York Infantry, Companies D and E
17th Pennsylvania Cavalry, Companies D and H

First Division

Brig. Gen. James Barnes

First Brigade	Second Brigade	Third Brigade
Col. William S. Tilton	Col. Jacob B. Sweitzer	Col. Strong Vincent
18th Massachusetts	9th Massachusetts	20th Maine
22nd Massachusetts	32nd Massachusetts	16th Michigan
1st Michigan	4th Michigan	44th New York
118th Pennsylvania	62nd Pennsylvania	83rd Pennsylvania

Second Division

Brig. Gen. Romeyn B. Ayres

First Brigade	Second Brigade	Third Brigade
Col. Hannibal Day	Col. Sidney Burbank	Brig. Gen. Stephen H. Weed
3rd United States	2nd United States	140th New York
4th United States	7th United States	146th New York
6th United States	10th United States	91st Pennsylvania
12th United States	11th United States	155th Pennsylvania
14th United States	17th United States	

Third Division

Brig. Gen. Samuel W. Crawford

First Brigade	Third Brigade
Col. William McCandless	Col. Joseph W. Fisher

1st Pennsylvania Reserves
2nd Pennsylvania Reserves
6th Pennsylvania Reserves
13th Pennsylvania Reserves

5th Pennsylvania Reserves
9th Pennsylvania Reserves
10th Pennsylvania Reserves
11th Pennsylvania Reserves
12th Pennsylvania Reserves

Fifth Corps Artillery

Capt. Augustus P. Martin

3rd Massachusetts Artillery (Battery C)
1st New York Artillery, Battery C
1st Ohio Artillery, Battery L
5th United States Artillery, Battery D
5th United States Artillery, Battery I

Sixth Army Corps

Maj. Gen. John Sedgwick

General Headquarters

1st Massachusetts Cavalry, temporarily attached
1st New Jersey Cavalry, Company L
1st Pennsylvania Cavalry, Company H

First Division

Brig. Gen. Horatio G. Wright

Provost Guard

4th New Jersey Infantry, Companies A, C, and H

First Brigade Brig. Gen. A.T.A. Torbert	Second Brigade Brig. Gen. Joseph J. Bartlett	Third Brigade Brig. Gen. David A. Russell
1st New Jersey	5th Maine	6th Maine
2nd New Jersey	121st New York	49th Pennsylvania
3rd New Jersey	95th Pennsylvania	119th Pennsylvania
15th New Jersey	96th Pennsylvania	5th Wisconsin

Second Division

Brig. Gen. Albion P. Howe

9

Second Brigade	Third Brigade
Col. Lewis A. Grant	Brig. Gen. Thomas H. Neill
2nd Vermont	7th Maine
3rd Vermont	33rd New York
4th Vermont	43rd New York
5th Vermont	49th New York
6th Vermont	77th New York
	61st Pennsylvania

Third Division

Maj. Gen. John Newton
Brig. Gen. Frank Wheaton

First Brigade	Second Brigade	Third Brigade
Brig. Gen. Alexander Shaler	Col. Henry L. Eustis	Brig. Gen. Frank Wheaton
65th New York	7th Massachusetts	62nd New York
67th New York	10th Massachusetts	93rd Pennsylvania
122nd New York	37th Massachusetts	98th Pennsylvania
23rd Pennsylvania	2nd Rhode Island	102nd Pennsylvania
82nd Pennsylvania		139th Pennsylvania

Sixth Corps Artillery

Col. Charles H. Tompkins
1st Massachusetts Artillery (Battery A)
1st New York Artillery, Independent Battery
3rd New York Artillery, Independent Battery
1st Rhode Island Artillery, Battery C
1st Rhode Island Artillery, Battery G
2nd United States Artillery, Battery D
2nd United States Artillery, Battery G
5th United States Artillery, Battery F

Eleventh Army Corps

Maj. Gen. Oliver O. Howard

General Headquarters

1st Indiana Cavalry, Companies I and K
8th New York Infantry
17th Pennsylvania Cavalry, Company K

First Division

Brig. Gen. Francis C. Barlow
Brig. Gen. Adelbert Ames

First Brigade
Col. Leopold Von Gilsa
41st New York
54th New York
68th New York
153rd Pennsylvania

Second Brigade
Brig. Gen. Adelbert Ames
17th Connecticut
25th Ohio
75th Ohio
107th Ohio

Second Division

Brig. Gen. Adolph Von Steinwehr

Provost Guard

29th New York Infantry, Independent Company

First Brigade
Col. Charles R. Coster
134th New York
154th New York
27th Pennsylvania
73rd Pennsylvania

Second Brigade
Col. Orland Smith
33rd Massachusetts
136th New York
55th Ohio
73rd Ohio

Third Division

Maj. Gen. Carl Schurz

First Brigade
Brig. Gen. Alexander
  Schimmelfennig
82nd Illinois
45th New York
74th Pennsylvania
157th New York
61st Ohio

Second Brigade
Col. Waldimir Krzyzanowski

58th New York
119th New York
26th Wisconsin
82nd Ohio
75th Pennsylvania

Eleventh Corps Artillery

Maj. Thomas W. Osborn

1st New York Artillery, Battery I
13th New York Artillery, Independent Battery
1st Ohio Artillery, Battery I
1st Ohio Artillery, Battery K
4th United States Artillery, Battery G

Twelfth Army Corps

Maj. Gen. Henry W. Slocum
Brig. Gen. Alpheus S. Williams

General Headquarters

9th New York Cavalry, Companies D and L

Provost Guard

10th Maine Infantry

First Division

Brig. Gen. Alpheus S. Williams
Brig. Gen. Thomas H. Ruger

First Brigade	Second Brigade	Third Brigade
Col. Archibald L. McDougall	Brig. Gen. Henry H. Lockwood	Brig. Gen. Thomas H. Ruger
5th Connecticut	1st Maryland Potomac	27th Indiana
20th Connecticut	Home Brigade	2nd Massachusetts
3rd Maryland	1st Maryland	13th New Jersey
123rd New York	Eastern Shore	107th New York
145th New York	150th New York	3rd Wisconsin
46th Pennsylvania		

Second Division

Brig. Gen. John W. Geary

Provost Guard

28th Pennsylvania Infantry, Company B

First Brigade	Second Brigade	Third Brigade
Col. Charles Candy	Col. George A. Cobham, Jr.	Brig. Gen. George S. Greene
	Brig. Gen. Thomas L. Kane	
5th Ohio	29th Pennsylvania	60th New York
7th Ohio	109th Pennsylvania	78th New York
29th Ohio	111th Pennsylvania	102nd New York
66th Ohio		137th New York
28th Pennsylvania		149th New York
147th Pennsylvania		

12

Twelfth Corps Artillery

Lt. Edward D. Muhlenberg
1st New York Artillery, Battery M
Pennsylvania Artillery, Independent Battery E
4th United States Artillery, Battery F
5th United States Artillery, Battery K

Cavalry Corps

Maj. Gen. Alfred Pleasonton

General Headquarters

1st Maine Cavalry, Company I

First Division

Brig. Gen. John Buford

First Brigade	Second Brigade	Reserve Brigade
Col. William Gamble	Col. Thomas C. Devin	Brig. Gen. Wesley Merritt

Provost Guard

8th Illinois	6th New York Cavalry,	6th Pennsylvania
12th Illinois	Company L	1st United States
3rd Indiana	6th New York	2nd United States
8th New York	9th New York	5th United States
	17th Pennsylvania	
	3rd West Virginia	

First Division Artillery

2nd United States Artillery, Battery A
1st United States Artillery, Battery K

Second Division

Brig. Gen. David McM. Gregg

Headquarters Guard

1st Ohio Cavalry, Company C

First Brigade	Third Brigade
Col. John B. Mcintosh	Col. J. Irvin Gregg
1st Maryland	1st Maine

Purnell (Maryland) Legion,
Company A
1st New Jersey
3rd Pennsylvania
3rd Pennsylvania Heavy Artillery

10th New York
4th Pennsylvania
16th Pennsylvania

Second Division Artillery

1st United States Artillery, Batteries E and G

Third Division

Brig. Gen. Judson Kilpatrick

Headquarters Guard

1st Ohio Cavalry, Company A

First Brigade
Brig. Gen. Elon J. Farnsworth
5th New York
18th Pennsylvania
1st Vermont
1st West Virginia

Second Brigade
Brig. Gen. George A. Custer
1st Michigan
5th Michigan
6th Michigan
7th Michigan

Third Division Artillery

2nd United States Artillery, Battery M
4th United States Artillery, Battery E

Cavalry Reserve Artillery

Capt. James M. Robertson
9th Michigan Artillery
6th New York Artillery, Independent Battery
2nd United States Artillery, Batteries B and L

Army Artillery Reserve

Brig. Gen. Robert O. Tyler
Capt. James M. Robertson

Headquarters Guard

32nd Massachusetts Infantry, Company C

Ammunition Train Guard

4th New Jersey Infantry (7 Companies)

First Regular Brigade
Capt Dunbar R. Ransom
1st United States Artillery,
  Battery H
3rd United States Artillery
  Batteries F and K
4th United States Artillery,
  Battery C
5th United States Artillery,
  Battery C

First Volunteer Brigade
Lt. Col. Freeman McGilvery
5 Massachusetts Artillery
  (Battery E)
9th Massachusetts Artillery
15th New York Artillery
  Independent Battery
Pennsylvania Artillery
  Independent
  Batteries C and F

Second Volunteer Brigade
Capt. Elijah D. Taft
2nd Connecticut Artillery,
  Independent Battery
5th New York Artillery,
  Independent Battery

Third Volunteer Brigade
Capt. James F. Huntington
1st New Hampshire
  Artillery, Battery A
1st Ohio Artillery, Battery H
1st Pennsylvania Artillery,
  Batteries F and G
1st West Virginia Artillery, Battery C

Fourth Volunteer Brigade
Capt. Robert H. Fitzhugh
6th Maine Artillery, (Battery F)
1st Maryland Artillery, Battery A
1st New Jersey Artillery, Battery A
1st New York Artillery, Battery G
1st New York Artillery, Battery K

5th CONNECTICUT INFANTRY
12th Corps, 1st Div., 1st Brig.

Raised: Counties of Fairfield, Hartford, New Haven, New London, Litchfield and Windham.

Organized: Camp Putnam, Hartford, Conn. M.I. July 22, 1861.

Commander: Col Warren Wightman Packer. b. Groton, Conn., Feb. 1, 1835. Clerk in Groton. Capt., Co. G, 5th Conn., July 23, 1861. Wd. Aug. 9, 1862. M.O. Oct. 20, 1864. d. Mystic, Conn., July 12, 1912.

Number: 324 **Loss:** 2 wd., 5 m.

Monument: Slocum Ave. Map reference III 7-G
"July 2, & 3, 1863"

14th CONNECTICUT INFANTRY
2nd Corps, 3rd Div., 2nd Brig.

Raised: Counties of New London, Tolland, Hartford, Middlesex, New Haven and Fairfield.

Organized: Camp Foote, near Hartford, Conn. M.I. Aug. 1862.

Commander: Maj. Theodore Grenville Ellis. b. Boston, Mass., Sept. 25, 1829. Civil engineer in Hartford. 1st Lt. and Adjt., 14th Conn., Aug. 23, 1862. M.O. Col., June 23, 1865. Surveyor-General of Conn. d. Hartford, Jan. 8, 1883.

Number: 200 **Loss:** 10 k., 52 wd., 4 m.

Monument: Hancock Ave. Map reference II F-11
"The 14th C.V. reached the vicinity of Gettysburg at evening July 1st, 1863, and held this position July 2nd, 3rd, and 4th. The regt. took part in the repulse of Longstreet's grand charge on the 3rd, capturing in their immediate front more than 200 prisoners and five battle-flags. They also, on the 3rd, captured from the enemy's sharp-shooters the Bliss buildings in their far front, and held them until ordered to burn them."

Markers: Two markers at the Bliss house site. II A-8, II A-9

17th CONNECTICUT INFANTRY
11th Corps, 1st Div., 2nd Brig.

Raised: Fairfield County

Organized: Camp Aiken, Seaside Park, Bridgeport, Conn. M.I. Aug. 28,1862.

Commanders: Lt. Col. Douglas Fowler. b. Guilford, Conn., Jan. 26,1826. In 1850 he was a Locksmith. Lived in Norwalk, Conn. at start of war. Capt., 3rd Conn Inf. (3 mos.). Capt., 8th Conn. Inf. until M.O. Feb. 1862. Capt., Co. A, 17th Conn., Aug. 28, 1862. Killed July 1, 1863 at Gettysburg. Maj. Allen G Brady then took command. b. Middlesex County, Mass., Feb. 13, 1822. Shirt manufacturer in Torrington, Conn. Lt. Col., 3rd Conn. Inf. (3 mos.). Capt, Co. B, 17th Conn., Aug. 28, 1862. M.O. Oct. 21, 1863. Maj., Veteran Reserve Corps. Lt., 43rd U.S. Inf., July 28, 1866. M.O. Mar. 4, 1867. d. Fayetteville, N.C., Feb. 11, 1905.

Number: 386   **Loss:** 20 k., 81 wd., 96 m.

Monuments: Howard Ave. and Wainwright Ave.

Map references **I C-15, II L-2**

Wainwright Ave. — "After a fierce contest with Early's Division at Barlow's knoll on July 1st, marked by monument there, this regiment formed in the line of battle on East Cemetery Hill and on the evening of July 2nd, took position here and was engaged in repulsing the desperate night assault of Haye's [Hays'] and Hokes [Hoke's] Brigades."

20th CONNECTICUT INFANTRY
12th Corps., 1st Div., 1st Brig.

Raised: Counties of New Haven, Hartford, and Middlesex.

Organized: Camp at Oyster Point, New Haven, Conn. M.I. Sept. 8, 1862.

Commander: Lt. Col. William Burr Wooster. b. Oxford, Conn., Aug. 22, 1821. Grad. Yale College Law School, 1846. Militia officer. Elected to state legislature. Lawyer in West Haven, Conn. Lt. Col., 20th Conn., Sept. 8, 1862. M.O. Mar. 7, 1864. Col., 29th Conn. Inf., Mar. 8, 1864. M.O. Aug. 21, 1865. d. Ansonia, Conn., Sept. 20, 1900.

Number: 434   **Loss:** 5 k., 22 wd., 1 m.

Monument: Slocum Ave. Map reference **III 7-G**

"The Brig. formed this line on the morning of July 2nd. At eve it moved to support left of army. Returning, it found the position and woods on rear occupied by Johnson's Division, Ewell's Corps. During the night it lay in line of battle. At dawn, July 3rd, the 20th Conn. advanced under cover of artillery and fought 5 hrs. driving the enemy and reoccupying the works. Was relieved by the 123rd N.Y. In afternoon moved to support the 2nd Corps against Longstreet's assault."

27th CONNECTICUT INFANTRY
2nd Corps., 1st Div., 4th Brig.

Raised: New Haven County

Organized: 9 month regiment organized at Camp Terry, New Haven, Conn. M.I. Oct. 22, 1862.

Commanders: Lt. Col. Henry Czar Merwin. b. Brookfield, Conn., Sept. 17, 1839. Militia service. Associated with family business in New Haven. Sergt., 2nd Conn. Inf. (3 mos.). Lt. Col., 27th Conn., Oct. 22, 1862. Killed July 2, 1863 at Gettysburg. When Merwin killed, Maj. James H. Coburn led regiment. b. probably in New Haven, Conn. Merchant in New Haven. Pvt., 2nd Conn. Inf. (3 mos.). Capt., Co. A, 27th Conn., Oct. 3, 1862. M.O. July 27, 1863. d. Bromley, Iowa, Dec. 20, 1899, age 63 years and 10 days old.

Number: Most of regiment captured at Chancellorsville, Va., May 1863. Only companies D & F survived to which was added a third company. 160 men in all.   **Loss:** 10 k., 23 wd., 4 m.

Monuments: Ayres and Brooke Avenues. Map references **V B-8, V C-5**

Sickles Ave. — "Charged over this ground the afternoon of July 2, 1863. The 4th Brigade forced the enemy from the Wheat Field and beyond the woods in front where the advanced position of the 27th Regt. is indicated by a tablet on the crest of the ledge. On this spot Lieut. Col. Merwin was killed while gallantly leading his command."

Markers: Two markers to Lt. Col. Merwin and where Capt. Jed Chapman fell. Map references **V B-8, V C-5**. Third marker indicates advanced position of regiment. Map reference **V C-5**

2nd CONNECTICUT ARTILLERY, INDEPENDENT BATTERY Artillery Reserve, 2nd Volunteer Brig.

Raised: Fairfield County.

Organized: Camp Buckingham, Seaside Park, Bridgeport, Conn. M.I. Sept. 10,1862.

Commander: Capt. John William Sterling. b. Bridgeport, Conn., Sept. 16, 1826. Merchant in Bridgeport. Capt., 2nd Conn. Artillery, July 22, 1862. M.O. Oct. 18, 1864. d. Bridgeport, June 5, 1881.

Number: 4 6-pdr. James Rifles, 2 12-pdr. Howitzers. 106 men.   **Loss:** 3 wd., 2 m.

Monument: Hancock Ave. Map reference **IV I-13**
"Position July 3, 1863."

Major General George G. Meade with 10th NY Infantry to the left rear.

18

1st DELAWARE INFANTRY
2nd Corps, 3rd Div., 2nd Brig.

Raised: Counties of New Castle, Kent and Sussex.

Organized: Camp Andrews, near Hares Corner, New Castle County, Del. M.I. Oct. 17, 1861.

Commanders: Lt. Col. Edward Paul Harris. b. Delaware, 1837 or 1838. Family from Sussex County. In 1860, Harris lived in Georgetown, Del. Capt., Co. E, 1st Del. Inf., Sept. 28, 1861. Wd. Dec. 13, 1862. M.O. Oct. 28, 1863. d. in 1890s. On July 2, Harris was put under arrest for withdrawing part of the line (restored to command on July 4). Capt. Thomas Bullock Hizar took command from Harris. b. Brandywine "near Chadds Ford", Del., June 15, 1833. Builder and contractor in New Castle, Del. Lt., Co. I, 1st Del., Sept. 2, 1861. Wd. July 2, 1863 and Aug. 25, 1864. M.O. Oct. 15, 1864. d. Duluth, Minn., Sept. 11, 1894. When Hizar wd., Lt. William Smith led the regiment. b. New Castle, Del., 1839. Morocco leather dresser in Wilmington, Del. Pvt., 1st Del. Inf. (3 mos.). Sergt., Co. A, 1 st Del. Inf. (3 yrs.), Aug. 19,1861. Killed July 3, 1863 at Gettysburg. Next, Lt. John T. Dent took the regiment. b. Delaware County, Penn., Nov. 27, 1837. Wood Turner in Wilmington, Del. Pvt., 1st Del. Inf. (3 mos.). Sergt., Co. G, 1st Del. Inf. (3 yrs.), Aug. 21, 1861. Wd. May 6, 1864. M.O. Maj., July 12, 1865. d. Wilmington, Feb. 22, 1914.

Number: 288   **Loss:** 10 k., 54 wd., 13 m.

Monument: Hancock Ave. Map reference **II F-10**
   "July 2nd and 3rd, 1863"

Marker: Site of Bliss House. Map reference **II A-8**

2nd DELAWARE INFANTRY
2nd Corps, 1st Div., 4th Brig.

Raised: Counties of New Castle, Delaware and Cecil County, Maryland. Also Philadelphia, Penn.

Organized: Camp Brandywine near Wilmington, Del. M.I. June 12, 1861.

Commanders: Col. William P. Baily. b. Wilmington, Del., Dec. 5, 1825. Importing business in New York City before the war. Lt. Col., 2nd Del., July 13, 1861. Wd. Dec. 13, 1862. M.O. May 16, 1864. d. North Plainfield, N.J., Feb. 1, 1883. On July 3, 1863 the regiment was commanded by Capt. Charles Henry Christman. b. Philadelphia, Penn. Merchant in Philadelphia. Capt., Co. B, 2nd Del., June 1, 1861. M.O. July 1, 1864. d. New York City, Nov. 8, 1886, age 54 years, 5 months, 8 days.

Number: 280   **Loss:** 11 k., 61 wd., 12 m.

Monument: Brooke Ave. Map reference **V D-5**
   "July 2nd, 1863"

Marker: Hancock Ave. Map reference **II F-10**
   Position of skirmish line on July 3, 1863

82nd ILLINOIS INFANTRY
11th Corps, 3rd Div., 1st Brig.

Other Names: "2nd Hecker Regiment"
Raised: Cook County mostly. Also St. Clair County.
Organized: Camp Butler, Clear Lake, Springfield, Ill. M.I. Oct. 23, 1862.
Commander: Lt. Col. Edward Selig Salomon. b. Schleswig, Prussia, Dec. 25, 1836. Came to U.S. in 1854. Lawyer in Chicago, Ill. Lt., 24th Ill. Inf. Lt. Col., 82nd Ill., Sept. 26, 1862. M.O. June 9, 1865. Governor of Washington Territory, 1870-72. d. San Francisco, California, July 18, 1913.
Number: 347 **Loss:** 4 k., 19 wd., 89 m.
Monument: Howard Ave. Map reference **I D-12**
"First line of Battle, July 1, 1863. Moved in retreat to Cemetery Hill. Occupied the crest of Cemetery Hill, July 2 & 3, 1863. Participated in repulse of Ewell's Corps on night of July 2, 1863."

8th ILLINOIS CAVALRY
Cavalry Corps, 1st Div., 1st Brig.

Raised: Counties of Kane, DeKalb, Whiteside, DuPage, Cook, McHenry and Winnebago.
Organized: Camp Kane, St. Charles, Ill. M.I. Sept. 18, 1861.
Commander: Maj. John Lourie Beveridge. b. Greenwich, New York, July 6, 1824. Lawyer from Evanston, Ill. Capt., Co. F, 8th Ill. Cav., Sept. 18,1861. Promoted Maj. to date Sept. 18, 1861 M.O. Nov. 2, 1863. Col., 17th Ill. Cav., Jan. 28, 1864. M.O. Feb. 7, 1866. Sheriff of Cook County, Ill. U.S. House of Representatives, 1871-73. Governor of Ill., 1873-77. d. Hollywood, California, May 3, 1910.
Number: 491 **Loss:** 1 k., 5 wd., 1 m.
Monument: Reynolds Ave. Map reference **I H-4**
"First line of battle July 1, 1863. Occupied until relieved by 1st Corps. One squadron picketed ridge east of Marsh Creek and supported by another squadron met enemy's right advance. Lieut. Jones, Co. E, fired first shot as the enemy crossed Marsh Creek Bridge. On reforming line regiment took an advanced position on Hagerstown Road. Late in the day delayed enemy's advance by attacking his right flank, thereby aiding the infantry in withdrawing to Cemetery Hill. In the evening encamped on left flank. July 2, 1863 Buford's Division retired toward Westminster."
Marker: Route 30 at Marsh Creek, 3 miles northwest of Gettysburg, opposite Knoxlyn Road. No map reference.
"1st shot at Gettysburg July 1, 1863, 7:30 a.m."

12th ILLINOIS CAVALRY
Cavalry Corps, 1st Div., 1st Brig.

Raised: Counties of Cook and Kankakee.

Organized: Cities of Springfield and Chicago, Ill. M.I. Feb. 24,1862. All or part of at least five companies were at Gettysburg: A, E, F, H, I.

Commander: For the campaign both the 12th Ill. Cav. and the 3rd Ind. Cav. were under the command of Col. George Henry Chapman. See 3rd Indiana Cav. for sketch.

Number: 12th Ill. and 3rd Ind. combined had 588 men.   **Loss:** 4 k., 10 wd., 6 m.

Monument: Reynolds Ave. Map reference I G-5

"First line of battle July 1, 1863. Held until relieved by 1st Corps. One squadron picketing ridge east of Marsh Creek. Met enemy's left advance. Regiment retired to ridge on left rear; with brigade fought dismounted, repulsing attacks of the enemy; covered the withdrawal of line to Cemetery Hill; and in the evening took position on left flank of the army. July 2, 1863, Buford's Division retired toward Westminster."

7th New Jersey Infantry, Sickles Avenue

21

7th INDIANA INFANTRY
1st Corps, 1st Div., 2nd Brig.

Raised: Counties of Dearborn, Decatur, Hendricks, Marion, Johnson and Ohio.
Organized: Camp Morton, Indianapolis, Ind. M.I. Sept. 13, 1861.
Commander: Col. Ira Glanton Grover. b. Brownsville, Ind., Dec. 26, 1832. Grad. Asbury University, Indiana, 1856. Lawyer from Greensburg, Ind. Resigned from elected position in lower house of state legislature to join 7th Ind. Inf. (3 mos.). Capt., Co. E, 7th Ind. Inf. (3 yrs.), Sept. 13, 1861. Wd. June 8 or 9, 1862. M.O. Sept. 20, 1864. d. Greensburg, May 30, 1876.
Number: 437 on July 3, 1863.   **Loss:** 2 k., 5 wd., 3 m.
Monument: Slocum Ave. Map reference III E-2
  "July 1, 2, & 3"

14th INDIANA INFANTRY
2nd Corps, 3rd Div., 1st Brig.

Raised: Counties of Knox, Parke, Martin, Greene, Vanderburgh, Vigo, Owen, Vermillion and Monroe.
Organized: Camp Vigo, Terre Haute, Ind., M.I. June 7, 1861.
Commander: Col. John Coons. b. Knox County, Ind., 1828 or 1829. Lawyer from Vincennes, Ind. Capt., Co. G, 14th Ind., June 7, 1861. Wd. Sept. 17, 1862. Killed May 12, 1864 at Spotsylvania Court House, Va.
Number: 236 **Loss:** 6 k., 25 wd.
Monument: East Cemetery Hill. Map reference II K-3
  "On the evening of July 2nd 1863, a determined effort was made by Hay's [Hays] and Hoke's Brigades of Early's Division of Confederate troops to carry Cemetery Hill by storm. The Union troops supporting the Batteries occupying this ground were over whelmed and forced to retire. Wiedrich's Battery was captured and two of Ricketts' guns were spiked. Carroll's Brigade, then in position south-west of the Cemetery was sent to the rescue, advancing in double quick time through the Cemetery and across the Baltimore Pike. The men went in with a cheer, the 14th Indiana met the enemy among the guns on this ground where a hand to hand struggle ensued resulting in driving the enemy from the hill. On this spot Isaac Morris, the color bearer of the 14th Ind., was killed, and many others fell nearby. The regiment then took position along the stone fence at the base of the hill South-east from this point, facing the east, the right and left flanks being designated by stone markers, there placed, which position it held to the close of the great battle."
Markers: Flank markers show position of the regiment 8 p.m. July 2 to close of the battle. Map reference II L-4

19th INDIANA INFANTRY
1st Corps, 1st Div., 1st Brig.

Raised: Counties of Madison, Delaware, Marion, Wayne, Randolph, Elkhart, Johnson and Owen.

Organized: Camp Morton, near Indianapolis, Ind. M.I. July 29, 1861.

Commander: Col. Samuel J. Williams. b. Montgomery, Va., 1831. Farmer in Selma, Ind. Capt., Co. K, 19th Ind., July 29, 1861. Killed in Battle of the Wilderness, Va., May 6, 1864.

Number: 339 **Loss:** 27 k., 133 wd., 50 m.

Monument: Meredith Ave. Map reference **I H-2**
"July 1, 1863."

20th INDIANA INFANTRY
3rd Corps, 1st Div., 2nd Brig.

Raised: Counties of Miami, Lake, Marshall, Fountain, Laporte, Cass, Tippecanoe, Marion, Porter and White.

Organized: Camp Tippecanoe, Lafayette, Ind. M.I. July 22, 1861.

Commanders: Col. John Wheeler. b. New Milford, Conn., Feb. 6, 1825. Editor in Crown Point, Ind. Capt., Co. B, 20th Ind., July 22, 1861. Killed July 2, 1863. When Wheeler killed, Lt. Col. William Calvin Linton Taylor took command. b. Lafayette, Ind., May 22, 1836. Grad. Ind. Univ., 1855. Lawyer in Lafayette. Lt., Co. G, 20th Ind., July 22, 1861. M.O. Sept. 29, 1864. d. Lafayette, Feb. 18, 1901.

Number: 496 **Loss:** 32 k., 114 wd., 10 m.

Monument: Cross Ave. Map reference **V C-7**
"July 2nd 1863." "Col. John Wheeler killed near by."

27th INDIANA INFANTRY
12th Corps, 1st Div., 3rd Brig.

Raised: Counties of Putnam, Daviess, Johnson, Lawrence, Monroe, Morgan, Jennings and Dubois.

Organized: Camp Morris, White River, Indianapolis, Ind. M.I. Sept. 12, 1861.

Commanders: Col. Silas Colgrove. b. Woodhull, N.Y., May 24, 1816. Lawyer, member of Ind. legislature and prosecuting attorney for the state. Lived in Winchester, Ind. before the war. Capt., 8th Ind. Inf. Col., 27th Ind., Aug. 30, 1861. Wd. May 3, 1863 and July 20, 1864. M.O. Dec. 30, 1864. d. Lake Kerr, Florida, Jan. 14, 1907. When Colgrove took command of brigade on July 2, Lt. Col. John Raush Fesler took command of the regiment. b. near Springfield, Ohio, Nov. 16, 1835. Merchant in Morgantown, Ind. Capt., Co. G, 27th Ind., Sept. 12, 1861. M.O. Nov. 11, 1864. d. Indianapolis, Feb. 19, 1920.

Number: 339 **Loss:** 23 k., 86 wd., 1 m.

Monument: Carman Ave. Map reference **III I-10**
"This monument marks the ground over which the left wing of the 27th Indiana advanced in a charge made by the regiment on the morning of July 3rd 1863."

Markers: Near Spangler's Spring indicating farthest advance, 6 a.m., July 3, 1863. Also markers Carman Ave., locating primary position. Map references **III I-9, III I-10**

1st INDIANA CAVALRY
2 Companies, I & K
11th Corps Escorts and Guards

Other Names: I-"Stewart's Independent Company"
  K-"Bracken's Independent Company"
Raised: I-Vigo County. K-Marion County.
Organized: Indianapolis, Ind. As independent companies assigned to 1st Ind.
  Cavalry, Aug. 20, 1861.
Commander: Capt. Abram Sharra. b. Westmoreland County, Pennsylvania.
  Laborer on the railroad in Terre Haute, Ind. Sergt., Stewart's Independent
  Co., July 4, 1861. Lt. Col., 11th Ind. Cav., July 16, 1864. M.O. Col., Sept. 19,
  1865. d. Evansville, Ind., Oct. 20, 1893, age 53 years, 8 months, 18 days.
Number: 52   **Loss:** 3 m.
Monument: None

3rd INDIANA CAVALRY
6 Companies, ABCDEF
Cavalry Corps, 1st Div., 1st Brig.

Other Names: "Forty Fifth Indiana Volunteers"
Raised: Counties of Switzerland, Harrison, Dearborn, Jefferson and Fayette.
Organized: 5 companies formed at North Madison, Ind. F company formed at
  Indianapolis, Ind. All M.I. July-Aug. 1861.
Commander: Col. George Henry Chapman. b. Holland, Mass., Nov. 22, 1832.
  Served in U.S. Navy and then published newspaper. Lawyer in Indianapolis
  at the start of the war. Maj., 3rd Indiana Cav., Nov. 2, 1861. M.O. Brig. Gen.,
  Jan. 7, 1866. Served as judge in Marion County and as member of state
  legislature. d. near Indianapolis, July 16, 1882.
Number: 369   **Loss:** 6 k., 21 wd., 5 m.
Monument: Reynolds Ave. Map reference **I G-5**
  "July 1, 1863."

3rd MAINE INFANTRY
3rd Corps, 1st Div., 2nd Brig.

Raised: Counties of Kennebec, Sagadahoc and Somerset.

Organized: Camp Hamlin, near the capital building, Augusta, Me. M.I. June 4, 1861.

Commanders: Col. Moses B. Lakeman. b. Boston, Mass., Sept. 11, 1828. At enlistment he was a provision dealer in Augusta. Capt., Co. I, 3rd Me., June 3, 1861. Wd. May 23, 1864. M.O. June 28, 1864. d. Malden, Mass., Mar. 17, 1907. On July 3, when Lakeman promoted to greater command, Capt. William C. Morgan commanded regiment. b. Liverpool, England, 1830 or 1831. Served in English army. Came to U.S. in 1851. Printer for a newspaper in Boston at the start of the war. Capt., Co. F, 3rd Me. Sept. 11, 1861. Maj., Dec. 8, 1863. Wd. May 6, 1864. Mort. wd. May 23, 1864. d. May 24, 1864. Buried Fredericksburg National Cemetery.

Number: 266 **Loss:** 18 k., 59 wd., 45 m.

Monument: Peach Orchard. Map reference IV L-1

"Detached from the brigade, fought here in the afternoon of July 2nd 1863, having been engaged in the forenoon at point in advance as indicated by a marker. July 3rd. In position on left centre of line, until afternoon, with other regiments of the brigade, it moved to support of the centre at time of the enemy's assault."

Markers: Hancock Ave. and Berdan Ave. Map references II F-12, and map IV inset.

4th MAINE INFANTRY
3rd Corps, 1st Div., 2nd Brig.

Raised: Counties of Knox, Waldo and Lincoln.

Organized: Camp Knox, Rockland, Me. M.I. June 15, 1861.

Commanders: Col. Elijah Walker. b. Union, Me., July 2, 1818. Lumber merchant in Rockland, Me. Capt., Co. B, 4th Me., June 14, 1861. Wd. July 2, 1863 and May 5, 1864. M.O. July 19, 1864. d. Somerville, Mass., Mar. 20, 1905. When Walker wounded, Capt. Edwin Libby commanded regiment. b. Thomaston, Me., Feb. 28, 1832. Sea captain, 1855-1861. Residence Rockland at enlistment. Sergt., Co. H, 4th Me., June 15, 1861. Wd. Dec. 13, 1862. Killed in Battle of the Wilderness, Va., May 5, 1864.

Number: 332 **Loss:** 11 k., 59 wd., 74 m.

Monument: Crawford Ave. Designed by Col. Walker. Position July 2, 1863. Map reference V F-8

Marker: Hancock Ave. Map reference II F-12

"July 3 in support here."

5th MAINE INFANTRY
6th Corps, 1st Div., 2nd Brig.

Raised: Counties of Cumberland, York, Androscoggin and Oxford.

Organized: Camp Preble, Cape Elizabeth, near Portland, Me. M.I. June 24, 1861.
Commander: Col. Clark Swett Edwards. b. Otisfield, Me., Mar. 26, 1824. Farmer in Bethel, Me. Capt., Co. I, 5th Me., June 25, 1861. M.O. July 27, 1864. d. Bethel, May 3, 1903.
Number: 340 **Loss:** None
Monument: Sedgwick Ave. Map reference **V C-12**
"Occupied this position from evening of July 2nd until close of battle."

6th MAINE INFANTRY
6th Corps, 1st Div., 3rd Brig.

Raised: Counties of Washington, Penobscot, Hancock and Piscataquis.
Organized: Cape Elizabeth, Portland, Me. M.I. July 15, 1861.
Commander: Col. Hiram Burnham. b. Narraguagus, Me., 1813 or 1814. Listed as day laborer, Narraguagus, in 1860. Lt. Col., 6th Me., July 15, 1861. Killed, Brig. Gen., at Chaffin's Farm, Va., Sept. 29, 1864.
Number: 439 **Loss:** None
Monument: Howe Ave. Map reference **V K-17**
"Held this position July 3, 1863. In afternoon moved to support of centre, then to Big Round Top."

7th MAINE INFANTRY
6 Companies, BCDFIK
6th Corps, 2nd Div., 3rd Brig.

Raised: Counties of Aroostook, Kennebec and Penobscot.
Organized: Augusta, Me. M.I. Aug. 21, 1861.
Commander: Lt. Col. Selden Connor. b. Fairfield, Me., Jan. 25, 1839. Grad. Tufts College, Mass., 1859. Law student in Woodstock, Vermont. Pvt., 1st Vt. Inf. (3 mos.). Lt. Col., 7th Me., Aug. 22, 1861. Col., 19th Me., Jan. 11, 1864. Wd. May 6, 1864. M.O. Brig. Gen., April 7, 1866. Elected Governor of Maine, 1876 d. Augusta, Me., July 9, 1917.
Number: 261 **Loss:** 6 wd.
Monument: Neill Ave. Designed by General Connor. Map reference **III M-16**
"July 3d 1863."

10th MAINE INFANTRY
3 Companies, ABD
12th Corps Provost Guard

Other Names: "Tenth Maine Battalion"
Raised: Counties of York and Cumberland.
Organized: Camp Preble, Cape Elizabeth, near Portland, Me. M.I. Oct. 4, 1861.
Commander: Capt. John Davis Beardsley. b. Woodstock, New Brunswick (Canada), Jan. 1, 1837. Militia officer. Owned and operated a sawmill in

Woodstock. "He brought his sawmill hands along and enlisted them in the 10th Me." Lt., Co. D, 10th Me., Oct. 4, 1861. Promoted Maj. of 109th Colored Inf., Sept. 7, 1864. M.O. Col., Mar. 6, 1865. d. Winslow, Arkansas, July 2, 1911.

Number: 170   **Loss:** None

Monument: Baltimore Pike. Map reference III E-12

Beardsley wrote in 1896: "The 10th Maine Battalion remained where it was during the entire battle, and I think it is true that they did not fire a shot at the enemy during those three days."

16th MAINE INFANTRY
1st Corps, 2nd Div., 1st Brig.

Raised: Counties of Kennebec, Somerset, Franklin, Oxford, Penobscot and Androscoggin.

Organized: Augusta, Me. M.I. Aug. 14, 1862.

Commanders: Col. Charles William Tilden. b. Castine, Me., May 7, 1832. Militia officer. Merchant in Castine. Lt., 2nd Me., May 28, 1861. Lt.. Col., 16th Me., July 9, 1862. Captured July 1 at Gettysburg. M.O. June 5, 1865. d. Hallowell, Me., Mar. 12, 1914. When Tilden captured, Capt. Daniel Marston commanded regiment. b. Phillips, Me., June 22, 1813. Grocery business in Phillips. Pvt., 9th Me. Infantry. Capt., Co. C, 16th Me., Aug. 14, 1862. Wd. Dec. 13, 1862. M.O. Dec. 22, 1864. Grocery business in Wisconsin. d. Phillips, Me., Nov. 26, 1891.

Number: 311   **Loss:** 9 k., 59 wd., 164 m.

Monument: Doubleday Ave. Map reference I E-7

"July 1st, 1863, fought here from 1 o'clock until 4 p.m. when the division was forced to retire, by command of Gen. Robinson to Col. Tilden, the regiment was moved to the right, near the Mummasburg road, as indicated by a marker there, with orders to 'hold the position at any cost.' July 2nd and 3rd, in position with the division on Cemetery Hill."

Marker: Near monument. Map reference I C-7

17th MAINE INFANTRY
3rd Corps, 1st Div., 3rd Brig.

Raised: Counties of Cumberland, Oxford, Franklin, York and Androscoggin.

Organized: Camp King, Trotting Park, Cape Elizabeth, Near Portland, Me. M.I. Aug. 18, 1862.

Commander: Lt. Col. Charles Benjamin Merrill. b. Portland, Me., April 14, 1827. Grad. Bowdoin College, Me., 1847, and Harvard University Law School, 1849. Militia service. Lawyer in Portland. Lt. Col., 17th Me., Aug. 1, 1862. M.O. Oct. 6, 1864. d. Portland, April 5, 1891.

Number: 392   **Loss:** 18 k., 112 wd., 3 m.

Monuments: De Trobriand Ave. and Hancock Ave.

Map references **V C-6, II E-17**

De Trobriand Ave. — "The Seventeenth Maine fought here in the Wheatfield 2 1/2 hours, and at this position from 4:10 to 5:45 p.m., July 2, 1863. On July 3, at the time of the enemy's assault, it reinforced the centre and supported artillery."

Hancock Ave. — Position of July 3, 1863.

19th MAINE INFANTRY
2nd Corps, 2nd Div., 1st Brig.

Raised: Counties of Knox, Waldo, Kennebec, Sagadahoc and Somerset.

Organized: Bath, Me. M.I. Aug. 25, 1862.

Commanders: Col. Francis Edward Heath. b. Belfast, Me., Feb. 28, 1838. Attnd. Waterville College, Me. Clerk in Waterville, Me. Lt., 3rd Me. Inf. Lt. Col., 19th Me., Aug. 25, 1862. Struck by piece of shell on July 3, 1863. M.O. Col., Nov. 4, 1863. Served in state legislature. d. Waterville, Dec. 20, 1897. Lt. Col. Henry Whitman Cunningham took command from Heath. b. Belfast, Me., Sept. 12, 1806. Militia officer. Ran "New England House" in Belfast. Capt., 4th Me. Infantry. Maj., 19th Me., Aug. 25, 1862. M.O. June 10, 1864. d. Washington, D.C., Oct. 7, 1871.

Number: 543   **Loss:** 29 k., 170 wd., 4 m.

Monument: Hancock Ave. Map reference II E-13

"In the evening of July 2d, this regiment at a position on the left of Batt'y G, 5th U.S., helped to repel the enemy that had driven in Humphreys' Division, taking one battle flag and re-capturing four guns. On July 3, after engaging the enemy's advance from this position, it moved to the right to the support of the 2d Brigade and joined in the final charge and repulse of Pickett's command."

Markers: Hancock Ave. Map reference II D-16

The flank markers, south of monument on Hancock Ave., locate the position of the 19th Me. on July 2, 1863 while supporting Weir's Battery, 5th U.S. Art.

20th MAINE INFANTRY
5th Corps, 1st Div., 3rd Brig.

Raised: Counties of Piscataquis, Penobscot, Lincoln and Knox.

Organized: Camp Mason, near Portland, Me. M.I. Aug. 29, 1862.

Commander: Col. Joshua Lawrence Chamberlain. b. Brewer, Me., Sept. 8, 1828. Grad. Bowdoin College, Me., 1852. Professor of modern languages at Bowdoin. Lt. Col., 20th Me., Aug. 8, 1862. Wounded six times during the war. Awarded Medal of Honor for actions at Gettysburg. M.O. Brig. Gen., Jan. 15, 1866. Governor of Me. President of Bowdoin. d. Portland, Me., Feb. 24, 1914.

Number: 386   **Loss:** 29 k., 91 wd., 5 m.

Monuments: 1. Little Round Top. Map reference V G-11

"Marks very nearly the spot where the colors stood." Stone lists those killed in battle.   2. Big Round Top. Map reference V I-9

Occupied this position on evening of July 2nd.

Marker: Location of Co. B, 200 yards S.E. of monument. Map reference **V G-12**. Not found.

Official Report of Chamberlain July 2, 1863.

"The line faced generally toward a more conspicuous eminence southwest of ours, which is known as Sugar Loaf, or Round Top. Between this and my position intervened a smooth and thinly wooded hollow. My line formed, I immediately detached Company B, Captain Morrill commanding, to extend from my left flank across this hollow as a line of skirmishers, with directions to act as occasion might dictate, to prevent a surprise on my exposed flank and rear.

"Mounting a large rock, I was able to see a considerable body of the enemy moving by the flank in rear of their line engaged, and passing from the direction of the foot of Great Round Top through the valley toward the front of my left.

"We opened a brisk fire at close range, which was so sudden and effective that they soon fell back among the rocks and low trees in the valley, only to burst forth again with a shout, and rapidly advanced, firing as they came. They pushed up to within a dozen yards of us before the terrible effectiveness of our fire compelled them to break and take shelter.

"They renewed the assault on our whole front, and for an hour the fighting was severe. Squads of the enemy broke through our line in several places, and the fight was literally hand to hand. The edge of the fight rolled backward and forward like a wave .... Forced from our position, we desperately recovered it, and pushed the enemy down to the foot of the slope.

"The enemy seemed to have gathered all their energies for their final assault. We had gotten our thin line into as good a shape as possible, when a strong force emerged from the scrub wood in the valley, as well as I could judge, in two lines in echelon by the right, and, opening a heavy fire, the first line came on as if they meant to sweep everything before them.

"My ammunition was soon exhausted. My men were firing their last shot and getting ready to 'club' their muskets.

"It was imperative to strike before we were struck by this overwhelming force in a hand-to-hand fight, which we could not probably have withstood or survived. At that crisis, I ordered the bayonet. The word was enough. It ran like fire along the line, from man to man, and rose into a shout, with which they sprang forward upon the enemy, now not 30 yards away. The effect was surprising; many of the enemy's first line threw down their arms and surrendered .... Holding fast by our right, and swinging forward our left, we made an extended 'right wheel,' before which the enemy's second line broke and fell back, fighting from tree to tree, many being captured, until we had swept the valley and cleared the front of nearly our entire brigade."

2nd MAINE ARTILLERY (BATTERY B)
1st Corps Artillery Brigade

Raised: Many counties represented. From Knox County primarily.
Organized: Augusta, Me. M.I. Nov. 30, 1861.
Commander: Capt. James Abram Hall. b. Jefferson, Me., Aug. 10, 1835. Merchant in Damariscotta, Me. Lt., 2nd Me. Art., Nov. 30, 1861. M.O. Lt. Col., 1st Battalion, Me. Art., July 22, 1865. Joined 2nd U.S. Veteran Volunteers. M.O. Col., Mar. 1, 1866. d. on a train in Syracuse, N.Y., June 10, 1893.
Number: 6 Ordnance Rifles. 127 men. **Loss:** 18 wd.
Monument: Chambersburg Pike (Rt. 30). I F-4
"July 1, 1863."
Marker: National Cemetery. Map reference II I-7
Position of July 2, 1863.

5th MAINE ARTILLERY (BATTERY E)
1st Corps Artillery Brigade

Raised: "At large."   Many men from Androscoggin County, Me. and Coos County, N.H.
Organized: Augusta, Me. M.I. Dec. 4, 1861.
Commanders: Capt. Greenlief Thurlow Stevens. b. Belgrade, Me., Aug. 20, 1831. Grad. Harvard University, 1861. Lawyer with residences in Belgrade and Augusta. Lt., 5th Me. Art., Jan. 31, 1862. Wd. July 2, 1863. M.O. July 6, 1865. Elected to state legislature. d. Augusta, Me., Dec. 21 1918. When Stevens wounded, Lt. Edward Newton Whittier commanded battery b. Gorham, Me., July 2, 1840. Grad. Brown University, 1862. Pvt., 1st Rhode Island Infantry (3 Mos.). Sergt., 5th Me. Art., Nov. 29, 1861. Awarded Medal of Honor for actions in the Battle of Fisher's Hill, Va. M.O. July 25, 1865. Grad. Harvard Medical School, 1869. d. Boston, June 14, 1902.
Number: 6 12-pdr. Napoleons. 136 men.   **Loss:** 3 k., 13 wd., 7 m.
Monument: Slocum Ave. Map reference III A-4
"Fought here July 1, 2, 3, 1863. Also engaged July 1st north of the Seminary (see marker). Ammunition expended, 979 rounds."
Marker: Seminary Ave. Map reference I I-6

6th MAINE ARTILLERY (BATTERY F)
Artillery Reserve, 4th Volunteer Brigade

Other Names: "McGilvery's"
Raised: Aroostook County
Organized: Augusta, Me. M.I. Jan. 1, 1862.
Commander: Lt. Edwin Barlow Dow. b. New Brunswick (Canada), June 20, 1835. Broker in New York City. Pvt., 8th New York Militia Inf. (3 mos.). Residence then Portland, Me., occupation, clerk. Lt., 6th Me. Art., Feb. 6, 1862. Wd. June 4, 1864. M.O. Capt., Nov. 29, 1864. d. New York City, June 29, 1917. Buried Arlington National Cemetery.

Number: 4 12-pdr. Napoleons. 103 men.　　**Loss:** 13 wd.
Monument: Hancock Ave. Map reference **IV J-13**
　　Fought here July 2, 1863.

1st MAINE CAVALRY
Cavalry Corps, 2nd Div., 3rd Brig.
Company L, 1st Corps Headquarters
Company I, Cavalry Corps Headquarters

Raised: Counties of Penobscot, Kennebec, Cumberland, Somerset, York, Aroostook and Franklin.
Organized: Camp Penobscot, fair grounds, Augusta, Me. M.I. Oct.-Nov., 1861.
Commander: Lt. Col. Charles Henry Smith. b. Hollis, Me., Nov. 1, 1827. Grad. Waterville College, Me. Teacher and law student in Eastport, Maine. Capt., Co. D, 1st Me. Cav., Oct. 19, 1861. Wd. June 24, 1864 at Saint Mary's Church. Awarded Medal of Honor for gallantry in that fight. M.O. Col., Aug. 1, 1865. Col., 28th U.S. Infantry and 19th U.S. Infantry. Retired Nov. 1,1891. d. Washington, D.C., July 17, 1902. Buried Arlington National Cemetery.
Number: 438　**Loss:** 1 k., 4 wd.
Monument: Hanover Road. No map reference.

20th Maine Infantry, Little Round Top.

31

1st MARYLAND EASTERN SHORE INFANTRY
12th Corps, 1st Div., 2nd Brig.

Raised: Baltimore City. Counties of Dorchester, Caroline, Talbot and Somerset.
Organized: Camp Wallace, Cambridge, Md. M.I. Sept., 1861.
Commander: Col. James Wallace. b. Dorchester County, Mar. 18, 1818. Grad. Dickinson College, Penn., 1840. Lawyer, militia officer and member of state legislature. Residence in Cambridge. Col., 1st Md. East. Shore, Aug. 16, 1861. He was a slaveowner who resigned, in part, over the question of enlistment of blacks for the Union cause. M.O. Dec. 23, 1863. d. Cambridge, Feb. 12, 1887.
Number: 583 **Loss:** 5 k., 18 wd., 2 m.
Monument: Slocum Ave. Map reference III F-4
"Five companies held the works in front of this stone wall on the morning of July 3, 1863, relieving other troops and remaining until about noon when they were relieved. The remainder of the regiment were in position during the same time about three hundred yards to the right."

1st MARYLAND POTOMAC HOME BRIGADE INFANTRY
12th Corps, 1st Div., 2nd Brig.

Raised: Baltimore City. Counties of Frederick and Washington.
Organized: Camp located north of Frederick City, Md. M.I. Aug.Dec.,1861.
Commander: Col. William Pinckney Maulsby, Sr. b. Bel Air, Md., July 10, 1815. Grad. Union College, N.Y. Lawyer in Frederick. Col., 1st Md. Pot. Home Brig., Nov. 29, 1861. Resigned Aug. 25, 1864. Judge in Frederick and member of Maryland Court of Appeals. d. Westminster, Md., Oct. 3, 1894.
Number: 739 **Loss:** 23 k., 80 wd., 1 m.
Monument: Slocum Ave, near Spangler's Spring. Map reference III H-8
"July 2nd reinforced the left wing between 5 and 6 o'clock p.m., charging under the immediate direction of Gen. Meade and recapturing three pieces of artillery. July 3rd, engaged the enemy at this point from 5 to 6 o'clock, a.m. At 11 a.m. went to the assistance of the 2nd Div., 12th Corps, engaging the enemy there for about four hours."

3rd MARYLAND INFANTRY
12th Corps, 1st Div., 1st Brig.

Raised: Baltimore City. Counties of Washington and Talbot in Maryland, and Preston County, West Virginia.
Organized: Formed in Baltimore City and Williamsport, Md. in early 1862. Company F, made up of 9 month drafted men, was organized at Easton, Md., and assigned to the regiment in Feb., 1863.

Commander: Col. Joseph M. Sudsburg. b. Bavaria, Mar. 17, 1827. Attnd. Austrian military school leading to service in Austrian army. Served also with Polish revolutionaries, the French army in Africa, and with Baden patriots against the Monarchy. Fled to Switzerland and in 1851 to U.S. Wood carver in Baltimore. Capt., 2nd Maryland Inf. Transferred to 4th Maryland Inf. Lt. Col., 3rd Md., May 7, 1862. M.O. June 24, 1864. d. Baltimore, April 8, 1901. Buried Loudon Park National Cemetery.

Number: 278 **Loss:** 1 k., 7 wd.

Monument: Slocum Ave. Map reference III G-7

"July 2nd, 1863, occupied this position in reserve. Late in the afternoon moved to reinforce the left of the line. Returning about 9 p.m. and finding the works occupied by the enemy. July 3rd. Under fire in reserve until about noon, then occupied the works in front and held them until relieved."

1st MARYLAND ARTILLERY, BATTERY A
Artillery Reserve, 4th Volunteer Brigade

Raised: Baltimore City

Organized: Formed as part of the Purnell Legion at the Pikesville, Md. arsenal in Aug. 1861.

Commander: Capt. James H. Rigby. b. Baltimore City, June 4, 1832. Militia service. Carpenter in Baltimore. Lt., Battery A, Aug. 20, 1861. M.O. Mar. 11, 1865. d. Aug. 5, 1889. Buried Loudon Park National Cemetery.

Number: 6 Ordnance Rifles. 107 men. **Loss:** None

Monument: Powers Hill. Map reference III F-18

"Occupied this position on the morning of July 2nd 1863, and remained in battery until the termination of the Battle, engaging a battery of the enemy on the 2nd, and on the morning of the 3rd, shelling the woods in front for nearly three hours, assisting in driving out the enemy."

PURNELL LEGION MARYLAND CAVALRY, COMPANY A
Cavalry Corps, 2nd Div., 1st Brig.

Raised: Harford County

Organized: Pikesville, Md., arsenal. M.I. Sept.-Nov., 1861.

Commander: Capt. Robert Emmet Duvall. b. Maryland, Sept. 17, 1812. Elected to state legislature. Farmer in Bel Air, Maryland. Capt., Co. A, Purnell Legion, Dec. 7, 1861. M.O. Jan. 26, 1865. d. Vale, Md., April 17, 1890.

Number: 78 **Loss:** None

Monument: Gregg Ave., East Cavalry Battlefield. No map reference.

"This detached Company, commanded by Capt. Robert E. Duvall, served in the cavalry engagement on the flank, July 2nd and 3rd, 1863."

1st MARYLAND CAVALRY
11 Companies
Cavalry Corps, 2nd Div., 1st Brig.

Raised: Baltimore City, Counties of Allegany and Washington in Maryland, Allegheny County in Penn., and Washington, D.C.

Organized: Baltimore City, Md. M.I. Nov. 1, 1861. Company M added after Gettysburg.

Commander: Lt. Col. James Monroe Deems. b. Baltimore, Jan. 5, 1818. Studied music in Baltimore and Germany. 1848 appointed instructor of music at University of Virginia. Teacher and composer of music in Baltimore when the war began. Maj., 1st Md. Cav., Dec. 20, 1861. M.O. Nov. 10, 1863. Wrote music for piano and voice. d. Baltimore, April 18, 1901.

Number: 335 **Loss:** 2 wd., 1 m.

Monument: Gregg Ave., East Cavalry Battlefield. No map reference.

11th Massachusetts Infantry, Emmitsburg Road.

1st MASSACHUSETTS INFANTRY
3rd Corps, 2nd Div., 1st Brig.

Raised: Counties of Suffolk and Norfolk.

Organized: Fanueil Hall, Boston, Mass. M.I. May 23-27, 1861.

Commander: Lt. Col. Clark B. Baldwin. b. Westminster, Vermont. Merchant in Boston. Capt., Co. E, 1st Mass., May 25, 1861. M.O. Nov. 10, 1864. 3 d. Melrose, Mass., Nov. 10, 1890, age 71 years, 11 months 10 days.

Number: 384 **Loss:** 16 k., 83 wd., 21 m.

Monument: Emmitsburg Road. Map reference **IV C-6**

"On July 2, 1863 from 11 a.m. until 6:30 p.m. the First Regiment Massachusetts Volunteer Infantry, Lieut. Col. Clark B. Baldwin commanding, occupied this spot in support of its skirmish line 800 feet in advance. The regiment subsequently took position in the brigade line and was engaged until the close of the action."

Marker: Marks skirmish line west of monument. Map reference **IV A-5**

2nd MASSACHUSETTS INFANTRY
12th Corps, 1st Div., 3rd Brig.

Raised: Counties of Middlesex, Essex, Suffolk, Norfolk and Worcester.

Organized: Camp Andrew, West Roxbury, Mass. M.I. May 25, 1861.

Commanders: Lt. Col. Charles Redington Mudge. b. New York City, Oct. 22, 1839. Attnd. Harvard University. Manufacturer in Swamp Scott, Mass. On Nov. 16, 1862 he wrote to a school friend, "if you will just look back to that Sunday morning when you and I jumped out of our beds at the news of the capture of Ft. Sumpter-I fully made up my mind to fight; and when I say fight, I mean win or die. I do not wish to stop the thing half-way. I wish to establish the government upon a foundation of rock." Lt., Co. F, 2nd Mass., May 25, 1861. Wd., May 25, 1862. On the morning of July 3, Mudge received an order to lead his men down across Spangler's meadow to drive Confederates from captured Union breastworks at the foot of Culp's Hill. At first he questioned the messenger, "Are you sure that is the order?" "Yes." "Well," said he, "it is murder, but it's the order. Up, men, over the works! Forward, double quick!" In the charge a bullet struck Mudge just below the throat and killed him. Maj. Charles Fessenden Morse now took command. b. Boston, Sept. 22, 1839. Grad. Cambridge Scientific School, 1857. Architect in Jamaica Plain, Mass. Lt., Co. B, 2nd Mass., May 25, 1861. Wd., Mar. 16, 1865. M.O. July 14, 1865. d. Boston Dec. 1, 1926.

Number: 401 **Loss:** 23 k., 109 wd., 4 m.

Monument: Carman Ave. The first regimental monument on the battlefield. Erected 1879. Lists members killed in the battle. Map reference **III H-10**

"From the hill behind this monument on the morning of July 3, 1863 the Second Mass. Inf. made an assault upon the Confederate troops in the works at the base of Culp's Hill opposite."

Markers: Tablets on Carman Ave. and Little Round Top not found. Stone markers near monument indicate primary position of regiment.

35

7th MASSACHUSETTS INFANTRY
6th Corps, 3rd Div., 2nd Brig.

Raised: Counties of Bristol, Plymouth and Norfolk.
Organized: Camp Old Colony, Taunton, Mass. M.I. June 15, 1861.
Commander: Lt. Col. Franklin P. Harlow. b. Springfield, Vermont, Dec. 8, 1827. Member of militia. Mechanic in South Abington (Whitman). Capt., Co. K, 7th Mass., June 15, 1861. Wd. May 3, 1863. M.O. June 27, 1864. d. Whitman, Mass., Nov. 28, 1905.
Number: 369 **Loss:** 6 wd.
Monument: Sedgwick Ave. Map reference **V C-13**
"This line was 100 yards to the rear."

9th MASSACHUSETTS INFANTRY
5th Corps, 1st Div., 2nd Brig.

Raised: Counties of Suffolk, Essex, Middlesex, Worcester and Norfolk.
Organized: Camp Wrightmen, on Long Island in Boston Harbor, Mass. M.I. June 11, 1861.
Commander: Col. Patrick Robert Guiney. b. Parkstown, Tipperary, Ireland, Jan. 15, 1835. Came to U.S. at age 7. Attnd. Holy Cross College, Mass. Lawyer in Roxbury, Mass. Capt., Co. D, 9th Mass., June 11, 1861. Wd. May 5, 1864 resulting in the loss of an eye. M.O, June 21, 1864. d. Boston, Mass., Mar. 21, 1877.
Number: 474 **Loss:** 1 k., 6 wd.
Monument: Sykes Ave. Map reference **V H-10**
"During the Battle of Gettysburg the Ninth Regt. was detached from the 2nd Brigade and held this position on Round Top."

10th MASSACHUSETTS INFANTRY
6th Corps, 3rd Div., 2nd Brig.

Raised: Counties of Berkshire, Hampden, Franklin and Hampshire.
Organized: Hampden Park, Springfield, Mass. M.I. June 21, 1861.
Commander: Lt. Col. Joseph Bailey Parsons. b. Northampton, Mass., April 29, 1828. Militia service. Farmer in Northampton, Mass. Capt., Co. C, 10th Mass., June 21, 1861. Wd. May 31, 1862. M.O. July 1, 1864. Chief of Police of Northampton d. Winthrop, Mass., June 4, 1906.
Number: 416 **Loss:** 4 wd., 5 m.
Monument: Sedgwick Ave. Map reference **V C-13**
"July 2nd, 1863."

11th MASSACHUSETTS INFANTRY
3rd Corps, 2nd Div., 1st Brig.

Raised: Counties of Suffolk, Middlesex and Essex.

Organized: Fort Warren, Boston, Mass. M.I. June 13, 1861.

Commander: Lt. Col. Porter D. Tripp. b. Kennebunk, Maine, June 21, 1826. Militia officer. Builder from Boston. Capt., Co. C, 11th Mass., June 13, 1861. M.O. June 24, 1864. d. Watertown, Mass., Nov. 6, 1873.

Number: 364   **Loss:** 23 k., 96 wd., 10 m.

Monument: Sickles Ave. and Emmitsburg Road. Map reference **IV D-6** "Upon this spot stood the 11th Mass. Regt. during the second days battle of Gettysburg, July 2, 1863."

12th MASSACHUSETTS INFANTRY
1st Corps, 2nd Div., 2nd Brig.

Other Names: "Webster Regiment." Named after Fletcher Webster, the first commander of the regiment and his father, the statesman Daniel Webster.

Raised: Counties of Suffolk, Plymouth, Norfolk and Essex.

Organized: Ft. Warren, Boston, Mass. M.I. June 26, 1861.

Commander: Col. James Lawrence Bates. b. Weymouth, Mass., Aug. 6, 1820. Traveled to California gold fields in 1849. Boot and shoe manufacturing business in Weymouth. Capt., Co. H, 12th Mass., June 26, 1861. Maj., 33rd Mass., Aug. 5, 1862. From Maj., 33rd Mass. to Col., 12th Mass., Sept. 9, 1862. Wd. slightly July 1, 1863, M.O. July 8, 1864. Banker in Boston. d. South Weymouth, Nov. 11, 1875. When Bates wd., Lt. Col. David Allen, Jr. led the regiment. b. Gloucester, Mass., Mar. 9, 1829. Carpenter in Gloucester. Capt., Co. K, 12th Mass., June 26, 1861. Wd. Sept. 17, 1862 and Dec. 13, 1862. Killed May 5, 1864 in the Battle of the Wilderness, Va.

Number: 301   **Loss:** 5 k., 52 wd., 62 m.

Monument: Doubleday Ave. Map reference **I C-7** "July 1, 1861."

Markers: Hancock Ave. and Ziegler's Grove. Map references **IV F-13, II G-7**

13th MASSACHUSETTS INFANTRY
1st Corps, 2nd Div., 1st Brig.

Raised: Counties of Suffolk, Middlesex, Norfolk and Worcester.

Organized: Fort Independence, Boston Harbor, Mass., July 16, 1861.

Commanders: Col. Samuel Haven Leonard. b. Bolton, Mass., July 10, 1825. Militia officer. Express business in Boston. Col., 13th Mass. Infantry, July 16, 1861. Wd. slightly July 1, 1863. M.O. Aug. 17, 1864. d. West Newton, Mass., Dec. 27, 1902. When Leonard wd., Lt. Col. Nathaniel Walter Batchelder assumed command of regiment. b. Grafton, Mass., Nov. 16, 1825. Member of militia. Dry goods business in Boston. Lt. Col., 13th Mass. Inf., July 16, 1861. Resigned and was discharged April 15, 1864, "For the good of the service." General Gouverneur K. Warren wrote, "getting rid of Lt. Col. Batchelder was a necessary preliminary to inducing his regiment to reenlist as veteran vol." d. Boston, June 21, 1868.

Number: 284   **Loss:** 7 k., 77 wd., 101 m.

Monument: Robinson Ave. Map reference **I D-8**

"July 1, 1863." Marks the site where the color bearer fell.

15th MASSACHUSETTS INFANTRY
2nd Corps, 2nd Div., 1st Brig.

Raised: Worcester County

Organized: Camp Scott, Worcester, Mass. M.I. July 12, 1861.

Commanders: Col. George Hull Ward. b. Worcester, Mass., April 26, 1826. Militia officer. Farmer in Worcester. Lt. Col., 15th Mass., July 24, 1861. Wd. Oct. 21, 1861. Left leg amputated. d. July 3, 1863 of wounds received July 2, 1863 at Gettysburg. When Ward wounded, Lt. Col. George Clesson Joslin took command. b. Leominster, Mass., Aug. 19, 1839. Militia officer. Salesman in Worcester. Capt., Co. I, 15th Mass., Aug. 5, 1861. Wd. Sept. 17, 1862 and Nov. 27, 1863. M.O. Aug. 9, 1864. d. Boston, Nov. 21, 1916.

Number: 304   **Loss:** 23 k., 97 wd., 28 m.

Monument: Hancock Ave. Map reference **II E-14**

"July 3, 1863."

Marker: Near Emmitsburg Road. Map reference **II C-13**

Stone locates where Col. Ward fell on July 2, 1863.

Additional metal sign on pole near Copse of Trees along Hancock Ave. indicates position of regiment in repulse of Pickett's Charge. No map reference.

16th MASSACHUSETTS INFANTRY
3rd Corps, 2nd Div., 1st Brig.

Raised: Middlesex County

Organized: Camp Cameron, Cambridge, Mass. M.I. June-July, 1861.

Commander: Lt. Col. Waldo Merriam. b. Mass., Feb. 23, 1839. Family from Boston. Militia service. Boston newspaper obituary notes, "when the war broke out [he was] traveling in Europe having just past his majority preparatory to entering one of the most respectable mercantile houses here as a partner." Lt. and Adjt., 16th Mass., Aug. 5, 1861. Wd. June 30, 1862 and July 2, 1863. Killed at Spotsylvania Court House, Va., May 12, 1864. When Merriam wounded at Gettysburg, Capt. Matthew Donovan led the regiment b. Dublin, Ireland, May 10, 1830. Settled in Lowell, Mass., 1847. Painter in Lowell. Lt., Co. D, 16th Mass. Inf., July 12, 1861. Wd. May 12, 1864. M.O. July 27, 1864. d. Lowell, Dec. 29, 1876.

Number: 307   **Loss:** 15 k., 53 wd., 13 m.

Monument: Emmitsburg Road. Map reference **IV E-5**

18th MASSACHUSETTS INFANTRY
5th Corps, 1st Div., 1st Brig.

Raised: Counties of Norfolk, Bristol and Plymouth.

Organized: Camp Brigham, Readville, Mass. M.I. Aug. 24, 1861.

Commander: Col. Joseph Hayes. b. South Berwick, Maine, Sept. 4, 1835. Grad. Harvard University as part of the class of 1855. Real estate broker in Boston. Maj., 18th Mass. Inf., Aug. 24, 1861. Wd. May 5, 1864. M.O. Brig. Gen., Aug. 24, 1865. d. New York City, Aug. 19, 1912.

Number: 281   **Loss:** 1 k., 23 wd., 3 m.

Monument: Sickles Ave. Map reference **V B-5**
    "July 2, 1863."

Marker: Little Round Top. Not found.

19th MASSACHUSETTS INFANTRY
2nd Corps, 2nd Div., 3 Brig.

Raised: Counties of Essex, Suffolk and Middlesex.

Organized: Camp Schouler, Lynnfield, Mass. M.I. Aug. 28, 1861.

Commander: Col. Arthur Forrester Devereux. b. Salem, Mass., April 27, 1836. Attnd. West Point and Harvard University. Militia officer. Bookkeeper in Salem. Capt., 8th Mass. Inf. (3 mos.) Lt. Col., 19th Mass., Aug. 3, 1861. Wd. Sept. 17, 1862. M.O. Feb. 27, 1864. Served one term in Ohio legislature. d. Cincinnati, Oh., Feb. 13, 1906.

Number: 231   **Loss:** 9 k., 61 wd., 7 m.

Monument: Hancock Ave. Map reference **II F-13**
    "Stood here on the afternoon of July 3, 1863."

Marker: Metal sign on pole near Copse of Trees along Hancock Ave. indicates position of regiment in repulse of Pickett's Charge. No map reference.

20th MASSACHUSETTS INFANTRY
2nd Corps, 2nd Div., 3rd Brig.

Raised: Counties of Suffolk, Norfolk and Nantucket.

Organized: Camp Massasoit, Readville, Mass. M.I. July-Aug., 1861.

Commanders: Col. Paul Joseph Revere. b. Boston, Sept. 10, 1832. Grandson of Revolutionary War hero Paul Revere. Grad. Harvard University, 1852. Agent in Boston. Maj., 20th Mass. Inf., July 19, 1861. Wd. Sept. 17, 1862 and July 2, 1863. d. of wounds July 4, 1863. When Revere wounded, Lt. Col. George Nelson Macy led the regiment. b. Nantucket, Mass., Sept. 24, 1837. Clerk in Nantucket. Lt., Co. I, 20th Mass., Aug. 8, 1861. Wd. July 3, 1863. Left hand amputated. M.O. July 29, 1865. Shot himself dead accidentally at his Boston home Feb. 13, 1875. Third officer in command at Gettysburg was Capt. Henry Livermore Abbott. b. Lowell, Mass., Jan. 21, 1842. Grad. Harvard University, 1860. Student of law in Boston. 2 Lt., 20th Mass., Aug. 28, 1861. Wd. June 30, 1862. Killed May 6, 1864 during the Battle of the Wilderness, Va. Rank at death, Maj.

Number: 301.   **Loss:** 30 k., 94 wd., 3 m.

Monument: Hancock Ave. Map reference **II E-13**

Monument made from a large conglomerate pudding stone brought from Roxbury where the regiment was partially recruited and where it was a landmark on the town's playground. "This monument marks the position occupied by the Twentieth Massachusetts Infantry in line of battle July 2nd and 3rd 1863 until advanced to the front of the copse of trees on its immediate right to assist in repelling the charge of Longstreet's Corps."

Marker: Metal sign on pole near Copse of Trees along Hancock Ave. indicates position of regiment in repulse of Pickett's Charge. No map reference.

22nd MASSACHUSETTS INFANTRY
5th Corps, 1st Div., 1st Brig.

Other Names: "Henry Wilson's Regiment." Named for the regiment's first Col. and former Senator from Mass.

Raised: Counties of Suffolk, Middlesex, Bristol, Norfolk and Essex.

Organized: Camp Schouler, Lynnfield, Mass. M.I. Sept.-Oct., 1861. At Gettysburg the regiment had 11 companies including 2nd Company Massachusetts Sharpshooters, attached.

Commander: Lt. Col. Thomas Sherwin, Jr. b. Boston, July 11, 1839. Grad. Harvard Univ., 1860. Teacher in Bolton, Mass. Raised a company for 15th Mass. Inf. but it was disbanded. Lt. and adjt., 22nd Mass. Inf., Oct. 8, 1861. M.O. Oct. 17, 1864. d. Boston, Dec. 19, 1914. "Pioneer in telephone industry."

Number: 348   **Loss:** 3 k., 27 wd., 1 m.

Monument: Sickles Ave. Map reference **V B-5**
"July 2, 1863."

Marker: Near monument, dedicated to 2nd Co. Mass. Sharpshooters. Map reference **V B-5**. In addition, small metal sign to Sharpshooters is located on south slope, Little Round Top. Map reference **V G-11**

28th MASSACHUSETTS INFANTRY
2nd Corps, 1st Div., 2nd Brig.

Raised: Counties of Suffolk, Middlesex and Worcester.

Organized: Camp Cameron, Cambridge, Mass., Dec. 13, 1861. In Nov. 1862 the regiment was assigned to duty with the famed Irish Brigade.

Commander: Col. Richard. Byrnes. b. Cavan, Ireland, 1833 or 1834. Joined United States Cavalry, May 21, 1856. Also service in 17th U.S. Infantry and 5th U.S. Cavalry. Col., 28th Mass., Oct. 18, 1862. Residence during war in Jersey City, New Jersey. Wd. June 3, 1864. d. Washington, D.C., June 12, 1864.

Number: 265   **Loss:** 8 k., 57 wd., 35 m.

Monument: Sickles Ave. Map reference **V B-5**
"This Regt. went into battle July 2,1863. Erected by the survivors and friends of the Regt. to mark the spot where it fought in defense of the American Union."

Marker: Sedgwick Ave. Map reference-not found.

32nd MASSACHUSETTS INFANTRY
5th Corps, 1st Div., 2nd Brig.
Company C, Headquarters, Artillery Reserve

Raised: Counties of Middlesex, Plymouth, Essex and Suffolk.

Organized: Fort Warren, Boston, Mass. M.I. during the fall of 1861 through the spring of 1862.

Commander: Col. George Lincoln Prescott. b. Littleton, Mass., May 21, 1829. Lumber dealer in Concord, Mass. Capt., 5th Mass. Inf. (3 mos.). Capt., Co. B, 32nd Mass., Dec. 28, 1862. d. June 19, 1864 of wounds received June 18, 1864 near Petersburg, Va.

Number: 406 **Loss:** 13 k., 62 wd., 5 m.

Monument: Sickles Ave. Map reference **V B-5**

"Withstood an attack of the enemy about 5 o'clock p.m. July 2, 1863. Withdrawn from here, it fought again in the Wheatfield." Monument designed by a veteran of the regiment.

Markers: Location of field hospital along Sickles Ave. and position of Co. C, on the Taneytown Road. Map reference for field hospital **V B-6**. Marker for Co. C not found.

33rd MASSACHUSETTS INFANTRY
11th Corps, 2nd Div., 2nd Brig.

Raised: Counties of Middlesex and Bristol.

Organized: Camp Stanton, Lynnfield, Mass. M.I. June-July, 1862.

Commander: Col. Adin Ballou Underwood. b. Milford, Mass., May 19, 1828. Grad. Brown University, 1849. Attnd. Harvard University Law School. Lawyer in Newton, Mass. Capt., 2nd Mass. Inf. Lt. Col., 33rd Mass., July 28, 1862. Wd. Oct. 29, 1863. M.O. Brig. Gen., Aug. 24, 1865. d. Boston, Mass., Jan. 14, 1888.

Number: 562 **Loss:** 7 k., 38 wd.

Monument: Slocum Ave. Map reference **II M-5**

"Detached from the Second Brigade, Second Division, Eleventh Corps on July Second 1863. After supporting the batteries in action on Cemetery Hill, while in position in a line extending westward from near this spot, withstood and assisted in repulsing a charge of the enemy's infantry in its front." When on detached service, the regiment acted under command of Brig. Gen. Adelbert Ames, 11th Corps, 1st Div.

37th MASSACHUSETTS INFANTRY
6th Corps, 3rd Div., 2nd Brig.

Raised: Counties of Berkshire, Hampden and Hampshire.

Organized: Camp Briggs, Pittsfield, Mass. M.I. Aug.-Sept., 1862

Commander: Col. Oliver Edwards. b. Springfield, Mass., Jan. 30, 1835. Master mechanic in foundry in Warsaw, Illinois. Returned to Massachusetts at the

41

start of the war. Lt. and Adjt., 10th Mass. Inf., June 21, 1861. Col., 37th Mass., Sept. 4, 1862. M.O. Brig. Gen., Jan. 15, 1866. Mayor of Warsaw, Ill. d. Warsaw, April 28, 1904.

Number: 593   **Loss:** 2 k., 26 wd., 19 m.

Monument: Sedgwick Ave. Map reference **V B-13**

"July 2nd, 1863. This line was 100 yards to the right rear."

MASSACHUSETTS SHARPSHOOTERS, 1st Company
2nd Corps, 2nd Div.

Other Names: "Andrew Sharpshooters" **Raised:** Essex County.

Organized: Lynnfield, Mass. M.I. Aug. 1861.

Commander: Capt. William Plumer. b. New Market, New Hampshire, Nov. 29, 1823. Grad. Harvard University, 1845. Lawyer in Lexington, Mass. Capt., 1st Company Sharpshooters, Oct. 2, 1862. Wd. June 1863. Rode in an ambulance to Gettysburg. M.O. Sept. 26, 1863. d. Lexington, Sept. 4, 1896.

Number: 50   **Loss:** 2 k., 6 m.

Monument: Hancock Ave. Map reference **II F-11**

"Unattached Mass. Vol. in action July 3, 4, & 5, 1863 in different position."

Marker: Hancock Ave. Map reference **II F-8**

For information on 2nd Company Sharpshooters see 22nd Mass. Inf.

1st MASSACHUSETTS ARTILLERY (BATTERY A)
6th Corps Artillery Brigade

Raised: Suffolk County

Organized: Camp Cameron, North Cambridge, Mass. M.I. Aug. 28, 1861.

Commander: Capt. William Henry McCartney. b. Boston, Mass., July 11, 1834. Militia service. Lawyer in Boston. Service with 3 mos. Boston Artillery Battery. Lt., 1st Mass. Artillery, Aug. 28, 1861. M.O. Oct. 19, 1864. d. North Mtn., Sullivan County, Penn., May 11, 1894.

Number: 6 12-pdr. Napoleons. 145 men.   **Loss:** None.

Monument: National Cemetery. Map reference **II I-7**

"July 3, 1863."

3rd MASSACHUSETTS ARTILLERY (BATTERY C)
5th Corps Artillery Brigade

Other Names: "Martin's"

Raised: Suffolk County.

Organized: Camp Schouler, Lynnfield, Mass. M.I. Sept. 5, 1861.

Commander: Lt. Aaron Francis Walcott. b. Boston, Mass., July 19, 1836. Bookkeeper in Boston. Pvt., 2nd Mass. Artillery, July 31, 1861. Transferred to 3rd Mass. Artillery, Dec. 1, 1861. M.O. Sept. 16, 1864. d. Melrose, Mass., Dec. 11, 1907.

Number: 6 12-pdr. Napoleons. 124 men.   **Loss:** 6 wd.

Monument: Wheatfield Road. Map reference V B-10
"July 2, 1863."
Marker: Grant Ave. Map reference V K-14
"Position July 3rd at 3 p.m."

5th MASSACHUSETTS ARTILLERY (BATTERY E)
Artillery Reserve, 1st Volunteer Brigade

Raised: Counties of Bristol and Suffolk.
Organized: Camp Schouler, Lynnfield, Mass. M.I. Dec. 3, 1861.
Commander: Capt. Charles Appleton Phillips. b. Salem, Mass., Jan. 31, 1841. Attnd. Harvard University. Studied law in Boston. 2nd Lt., 5th Mass. Artillery, Oct. 23, 1861. M.O. June 12, 1865. Lawyer in Gold Hill, near Virginia City, Nevada where he died Mar. 20, 1876.
Number: 6 Ordnance Rifles. 104 men.   **Loss:** 2 k., 14 wd.
Monument: Wheatfield Road. Map reference IV L-3
"July 2, 1863."
Marker: Hancock Ave. Map reference IV G-13
"July 2. Withdrew at 5 p.m. from the field near the Peach Orchard and went into battery here. July 3. About 1:30, by order of Brig. General H. J. Hunt, fired on the Confederate batteries but did little damage. Opened an enfilading fire soon after on Longstreet's advancing line of infantry and assisted in repulsing the assault. A charge was made within the range of the battery immediately afterwards by the Florida brigade and at about the same time a Confederate battery opened on the left front which at once received the concentrated fire of the batteries of the brigade driving the cannoneers from their guns which they abandoned. July 4. Remained in this position until afternoon."

9th MASSACHUSETTS ARTILLERY
Artillery Reserve, 1st Volunteer Brigade

Raised: Middlesex County.
Organized: Camp Stanton, Lynnfield, Essex County, Mass. M.I. Aug. 10, 1862.
Commanders: Capt. John Bigelow. b. Brighton, Mass., Feb. 4, 1841. Grad. Harvard University, 1861. Service in 2nd Mass. Artillery and First Battalion, Maryland Artillery. Wd. July 1, 1862. Capt., 9th Mass. Artillery, Feb. 11, 1861. Wd. July 2, 1863. M.O. Dec. 16, 1864. d. Minneapolis, Minnesota, Sept. 13, 1917. When Bigelow wd., Lt. Richard Sweet Milton took command. b. Nov. 21, 1840. Bookkeeper in West Roxbury, Mass. Service with 4th Battalion, Mass. Inf. 2 Lt., 9th Battery, Aug. 10, 1862. M.O. Capt., June 6, 1865. d. Boston, Nov. 25, 1904.
Number: 6 12-pdr. Napoleons. 110 men.   **Loss:** 8 k., 18 wd., 2 m.
Monument: Wheatfield Road. Map reference IV M-5
"1st position left gun Wheatfield Road 4:30 to 6 p.m. July 2, 1863. Shelled Confederate Batteries on Emmitsburg Road also the enemy around Rose

Farm buildings. Enfiladed with canister Kershaw's Brigade C.S.A. moving across field in front from Emmitsburg Road to woods on left where battle was raging in front of Round Tops. 6 p.m.- alone on field. Graham's Brigade 3rd Corps forced from Peach Orchard had retired by detachments.

By 'prolonge firing' retired before Kershaw's skirmishers and Barksdale's Brigade C.S.A. 400 yards.

2nd position angle of stone wall near Trostle's House where the Battery was halted by Lieut. Colonel McGilvery and ordered to hold enemy in check until line of artillery could be formed 560 yards in the rear. Was without support and hemmed in by stone wall. Enemy closed in on flanks. Men and horses were shot down when finally overcome at 6:30 p.m. Lieut.-Colonel McGilvery had batteries unsupported in position near the Weikert House covering opening in lines between Round Tops and left of 2nd Corps 3/4 mile occasioned by withdrawal of Graham's Brigade.

7:15 p.m. Willard's Brigade 2nd Corps and later Lockwoods' Brigade 12th Corps came to support of artillery. 8 p.m. the enemy finally repulsed."

Markers: Hancock Ave. and near Trostle House. Map references **II F-9, IV K-7**

1st MASSACHUSETTS CAVALRY
8 Companies, ABCDEFGH
Temporarily attached to 6th Corps Headquarters for the Campaign.

Raised: Counties of Suffolk, Hampden and Essex.
Organized: Camp Brigham, Readville, Mass. M.I. Sept. 5 - Nov. 1, 1861.
Commander: Lt. Col. Greely Stevenson Curtis. b. Boston, Mass., Nov. 21, 1830. Attnd. Scientific School of Harvard University. Engineer in Boston. Capt., 2nd Mass. Inf., May 25, 1861. Resigned Oct. 28, 1861 to accept commission of Maj. in 1st Mass. Cav. M.O. Mar. 4, 1864. d. Boston, Feb. 12, 1897.
Number: 292   **Loss:** None
Monument: Sedgwick Ave. Map reference **V C-13**
"On Detached Service."

1st MICHIGAN INFANTRY
5th Corps, 1st Div., 1st Brig.

Raised: Counties of Washtenaw, Wayne and Jackson.

Organized: Camp Fountain, Fountain Street, Ann Arbor, Mich. M.I. Aug. 17, 1861.

Commanders: Col., Ira Coray Abbott. b. Burns, N.Y., Dec. 14, 1824. Grain dealer in Burr Oak, Mich. Capt., 1st Mich. (3 mos.). Capt., Co. B, 1st Mich. (3 yrs.), Aug. 17, 1861. Wd. Dec. 13, 1862 and July 2, 1863. M.O. Dec. 22, 1864. d. Washington, D.C., Oct. 9, 1908. Buried Arlington Nat'l Cem. When Abbott wounded, Lt. Col. William Alexander Throop took command. b. Schoharie, N.Y., July 26, 1838. Bookseller in Detroit, Mich. 2 Lt., 1st Mich. (3 mos.). Capt., Co. B, 1st Mich. (3 yrs.), Sept. 16, 1861. Wd. July 2, 1863, May 30, 1864 and July 30, 1864. M.O. Col., Jan. 6, 1865. d. by suicide, Detroit, Oct. 2, 1884.

Number: 261   **Loss:** 5 k., 33 wd., 4 m.

Monument: Sickles Ave. Map reference **V B-4**

"This monument marks the position where the regiment fought, July 2, 1863."

3rd MICHIGAN INFANTRY
3rd Corps, 1st Div., 3rd Brig.

Raised: Counties of Kent, Ionia, Ottawa, and Muskegon.

Organized: "Cantonment Anderson" on fair grounds south of city of Grand Rapids, Mich. M.I. June 10, 1861.

Commanders: Col. Byron Root Pierce. b. East Bloomfield, N.Y., Sept. 20, 1829. Dentist in Grand Rapids. Militia officer. Capt., Co. K, 3rd Mich., June 10, 1861. Wd. May 3, 1863 and July 2, 1863. Left leg amputated. M.O. Brig. Gen., April 24, 1865. d. Grand Rapids, July 10, 1924. When Col. Pierce was wounded, his brother Lt. Col. Edwin Sheldon Pierce led the regiment. b. East Bloomfield, N.Y., Dec. 3, 1831. Merchant in Grand Rapids. Capt., Co. E, 3rd Mich., May 13, 1861. Wd. May 3, 1863. M.O. Jan. 20, 1864. d. Aug. 31, 1912. Buried Arlington National Cemetery.

Number: 286   **Loss:** 7 k., 31 wd., 7 m.

Monument: Peach Orchard. Map reference **IV M-1**

"This regiment deployed as skirmishers 150 yards in advance of this position, held the line extending from the Peach Orchard east to the woods. Was the right of De Trobriand's Brigade, and connected with the left of Graham's."

4th MICHIGAN INFANTRY
5th Corps, 1st Div., 2nd Brig.

Raised: Counties of Lenawee, Hillsdale, Washtenaw and Monroe.

Organized: Camp Williams, Adrian College, Adrian, Mich. M.I. June 20, 1861.

Commanders: Col. Harrison H. Jeffords. b. Monroe County, N.Y., Aug. 21, 1834. Grad. Univ. of Mich. Law School, 1861. Lawyer in Dexter Mich. Lt., Co. K, 4th Mich., June 20, 1861. d. July 3, 1863 of wounds received on July 2.

"Thrust through with a bayonet while gallantly attempting to rescue his colors from the grasp of the enemy." When Jeffords wd., Lt. Col. George W. Lumbard assumed command of regiment. b. New York State, 1830. Lawyer in Hillsdale, Mich. Capt., Co. E, 4th Mich., May 16, 1861. d. Col., May 6, 1864 of wounds received on May 5, 1864 on the Wilderness, Va. battlefield.

Number: 403 **Loss:** 25 k., 64 wd., 76 m.

Monument: De Trobriand Ave. Map reference **V C-6**

"This monument marks the position held by the regiment July 2nd 1863." Also marks the location where Col. Jeffords fell.

5th MICHIGAN INFANTRY
3rd Corps, 1st Div., 3rd Brig.

Raised: Counties of St. Clair, Macomb, Saginaw, Oakland, Wayne, Livingston, and Shiawassee.

Organized: Fort Wayne, Detroit, Mich. M.I. Aug. 28, 1861.

Commander: Lt. Col. John Pulford. b. New York City, July 4, 1837. Hotel owner in Detroit. Lt., Co. A, 5th Mich., Aug. 28, 1861. M.O. Col., July 5, 1865. Wd., July 1, 1862, May 3, 1863, July 2, 1863, May 5, 1864 and Oct. 27, 1864. Entered Regular service until retirement on Dec. 15, 1870. d. Detroit, July 11, 1896.

Number: 283 **Loss:** 19 k., 86 wd., 4 m.

Monument: Sickles Ave. Map reference **V B-6**

"The regiment fought here about 4:30 o'clock p.m., July 2, 1863, after it had been assembled from the skirmish line far in advance of this position. It moved to the support of the 2nd Corps in resisting Pickett's charge, July 3."

7th MICHIGAN INFANTRY
2nd Corps, 2nd Div., 3rd Brig.

Raised: Counties of Monroe, Lapeer, Tuscola and Oakland.

Organized: Fair grounds in Monroe, Mich. M.I. Aug. 22, 1861.

Commanders: Lt. Col. Amos Steele, Jr. b. New York, 1834. Militia officer. Farmer in Mason, Mich. Lt., Co. B, 7th Mich., Oct. 31, 1861. Killed July 3, 1863 at Gettysburg. When Lt. Col. Steele was killed, Maj. Sylvanus Wright Curtiss led the regiment. b. Genesee County, N.Y., January 1, 1831. Butcher in Monroe, Mich. Lt., Co. D, 7th Mich., Aug. 22, 1861. Wd. Sept. 17, 1862 and May 1864. M.O. Oct. 5, 1864. Lived in Monroe after the war. d. Nov. 3, 1895.

Number: 165 **Loss:** 21 k., 44 wd.

Monument: Hancock Ave. Map reference **II E-13**

"Regiment held this position during the engagement of July 2nd and 3rd, 1863. On the evening of the 2nd changed front to the left, meeting and aiding in driving back the enemy. On the 3rd assisted in repulsing Pickett's charge, changing front to the right and assaulting the advancing force in flank."

16th MICHIGAN INFANTRY
5th Corps, 1st Div., 3rd Brig.

Raised: Counties of Wayne, Saginaw, Ionia and Ontonagon.
Organized: Camp Backus, Clinton Ave., Detroit, Mich. M.I. Sept. 8, 1861.
Commander: Lt. Col. Norval E. Welch. b. Pittsfield, Mich. around 1835. Private secretary to Senator Lewis Cass of Michigan. For a short period was appointed acting Governor of Nebraska Territory. Grad. University of Michigan Law School, 1860. Practiced law in Ann Arbor, Mich. Maj., 16th Mich., Aug. 22, 1861. Killed, as Col., in the Battle of Peebles' Farm, Va., Sept. 30, 1864.
Number: 356　**Loss:** 23 k., 34 wd., 3 m.
Monument: Sykes Ave. Map reference V F-10
"Regiment held this position during the afternoon and night of July 2, 1863, and assisted in defeating the desperate attempts of the enemy to capture Little Round Top."

24th MICHIGAN INFANTRY
1st Corps, 1st Div., 1st Brig.

Raised: Mostly Wayne County.
Organized: Camp Barns, fair grounds, Woodward Ave., Detroit, Mich., M.I. Aug. 15, 1862.
Commanders: Col. Henry Andrew Morrow. b. Warrenton, Va., July 10, 1829. Educated at Rittenhouse Academy in Washington, D.C. Serving as a Senate page, he came under the tutelage of Senator Lewis Cass of Michigan. Mexican War service. Lawyer in Detroit. Col., 24th Mich., July 15, 1862. Wd. July 1, 1863 and May 5, 1864. M.O. July 19, 1865. Regular service. d. Hot Springs, Arkansas, Jan. 31, 1891. When Morrow was wd., Capt. Albert Marshall Edwards led the regiment. b. Otisfield, Me., Feb. 25, 1836. Grad. Gould Academy, Me., 1856. Attnd. University of Michigan. Associate Editor in Chief of the Young Men's Journal and Temperance Advocate in Detroit. Sergt., 1st Mich. (3 mos.). Capt., Co. F, 24th Mich., June 9, 1862. M.O. Lt. Col., June 30, 1865. d. Detroit, July 15, 1909.
Number: 496　**Loss:** 67 k., 210 wd., 86 m.
Monument: Meredith Ave. Map reference I H-3
"Arriving upon the field to the south of these woods in the forenoon of July 1st, this regiment with others of the brigade charged across the stream in front to the crest beyond assisting in the capture of a large portion of Archer's Tennessee Brigade. It was then withdrawn to this position where it fought until the line was outflanked and forced back. Position July 2nd and 3rd on Culp's Hill."
Marker: Slocum Ave. Map reference III C-3

NINTH MICHIGAN BATTERY
(1st Michigan Artillery, Battery I)
Cavalry Corps, Reserve Artillery Brigade

Raised: Counties of Lenawee and Hillsdale

Organized: Detroit, Mich. as part of 5th Mich. Cavalry. M I. Aug. 29, 1862

Commander: Capt., Jabez J. Daniels. b. Hull, England, May 12, 1830. Came to Michigan as an infant. Merchant in Hudson, Mich. 2 Lt., 1st Mich. Cav., Aug. 22, 1861. Capt., 9th Mich. Battery, Aug. 29, 1862. Resigned Dec. 15, 1863. d. Hudson, Mich., Feb. 9, 1880.

Number: 6 Ordnance Rifles. 119 men.   **Loss:** 1 k., 4 wd.

Monument: Hancock Ave. Map reference II E-17

"This monument marks the position held by the Ninth Michigan Battery from 12:30 p.m., July 3rd, until 7 a.m. the following morning."

1st MICHIGAN CAVALRY
Cavalry Corps, 3rd Div., 2nd Brig.

Raised: Counties of Wayne, Lapeer and Oakland.

Organized: Camp Lyon at old race track in Hamtramck, Mich. M.I. Sept. 13, 1861.

Commander: Col. Charles H. Town. b. Elba, New York, May 28, 1828. Machinist in Detroit, Mich. Capt., Co. B, 1st Mich. Cav., Aug. 22, 1861. Wd. Aug. 30, 1862. M.O. Aug. 17, 1864. d. Elba, N.Y., May 7, 1865. "He always sought death on the battlefield, but never found it, and came home to die of consumption after the war was over."

Number: 502   **Loss:** 10 k., 43 wd., 20 m.

Monument: No unit monument. See brigade monument, East Cavalry Battlefield.

5th MICHIGAN CAVALRY
Cavalry Corps, 3rd Div., 2nd Brig.

Raised: Counties of Wayne, Oakland, Kalamazoo, Allegan and Branch.

Organized: Detroit, Mich. M.I. Aug. 30, 1862.

Commander: Col. Russell Alexander Alger. b. Lafayette Township, Medina County, Ohio, Feb. 27, 1836. Lumber business in Grand Rapids, Mich. Capt., 2nd Mich. Cav. Lt. Col., 6th Mich. Cav. Col., 5th Mich. Cav., June 11, 1863. Wd. July 1, 1862 and July 8, 1863. M.O. Sept. 20, 1864. Governor of Michigan, 1885-87. Secretary of War, 1897-99, and elected member of U.S. Senate, 1902-07. d. Washington, D.C., Jan. 24, 1907.

Number: 770   **Loss:** 8 k., 30 wd., 18 m.

Monument: No unit monument. See brigade monument, East Cavalry Battlefield.

6th MICHIGAN CAVALRY
Cavalry Corps, 3rd Div., 2nd Brig.

Raised: Counties of Kent, Ionia, Barry and Shiawassee.
Organized: Grand Rapids, Mich. M.I. Oct. 13, 1862.
Commander: Col. George Gray. b. County Tyrone, Ireland, June 20, 1824. Grad. Dublin University. City Attorney for Grand Rapids. Col., 6th Mich. Cav., Oct. 13, 1862. M.O. May 19, 1864. d. Orange, N.J., April 20, 1892.
Number: 611   **Loss:** 1 k., 26 wd., 1 m.
Monument: No unit monument. See brigade monument, East Cavalry Battlefield.

7th MICHIGAN CAVALRY
10 Companies, ABCDEFGHIK
Cavalry Corps, 3rd Div., 2nd Brig.

Raised: Counties of Saginaw, Eaton, Lenawee and Kalamazoo.
Organized: Lee Barracks in Grand Rapids, Mich. M.I. Jan. 27, 1863.
Commander: Col. William D'Alton Mann. b. Sandusky, Ohio, Sept. 27, 1839. Engineer and manufacturer in Detroit, Mich. Capt., 1st Mich. Cav. Lt. Col., 5th Mich. Cav. Col., 7th Mich. Cav., Feb. 19, 1863. M.O. Mar. 1, 1864. Inventor, politician and magazine owner. d. Morristown, N.J., May 17, 1920.
Number: 461   **Loss:** 13 k., 48 wd., 39 m.
Monument: No unit monument. See brigade monument, East Cavalry Battlefield.

National Cemetery with view of Major General John F. Reynolds, 75th Pennsylvania Infantry, and the Soldiers' National Monument.

1st MINNESOTA INFANTRY
2nd Corps, 2nd Div., 1st Brig.
Company C, Provost Guard, 2nd Div., 2nd Corp.

Raised: Counties of Ramsey, Hennepin, Washington, Dakota, Rice, Goodhue, Wabasha and Winona.

Organized: Fort Snelling, Minn. M.I. for three years to date from April 29, 1861. At Gettysburg the regiment had 11 companies, Co. L being the 2nd Company Minnesota Sharpshooters, attached.

Commanders: Col. William Colvill, Jr. b. Forestville, N.Y., April 5, 1830. Lawyer. Start of war was a newspaper editor in Red Wing, Minnesota. Capt., Co. F, 1st Minn., April 29, 1861. Wd. June 30, 1862 and July 2, 1863. M.O. Jan. 11, 1864. Member of Minnesota legislature. Attorney General of State, 1866-68. d. Minneapolis, Minnesota, June 13, 1905. When Colvill wd., Capt. Nathan S. Messick took command. b. New Jersey, 1827 or 1828. Family, I believe, from Gloucester County, N.J. Mexican War service. Shoemaker in Faribault, Minn. Lt., Co. G, 1st Minn., April 29, 1861. Killed July 3, 1863 at Gettysburg. Buried Gettysburg National Cemetery. Capt. Henry C. Coates led regiment after Messick. b. Philadelphia, Penn., 1832 or 1833. Printer in St. Paul, Minnesota. Lt., Co. A, 1st Minn., April 29, 1861. M.O. May 3, 1864. d. at residence in Wildwood, New Jersey, Sept. 18, 1909.

Number: 420. During the famous charge on July 2 at Gettysburg, Co. L (32 men) was serving as skirmishers and Co. C (56 men) was detached to the division.   **Loss:** 50 k., 173 wd., 1 m.

Monuments: Two on Hancock Ave. Map references **II E-14, IV F-13**
"On the afternoon of July 2, 1863 Sickles Third Corps having advanced from this line to the Emmitsburg road eight companies of the First Minnesota regiment numbering 262 men were sent to this place to support a battery. Upon Sickles repulse as his men were passing here in confused retreat two Confederate brigades in pursuit were crossing the swale. To gain time to bring up the reserves and save this position General Hancock in person ordered the eight companies to charge the rapidly advancing enemy. The order was instantly repeated by Col. Wm. Colvill and the charge instantly made down the slope at full speed through the concentrated fire of the two brigades breaking with the bayonet the enemy's front line as it was crossing the small brook in the low ground. There the remnant of the eight companies nearly surrounded by the enemy held its entire force at bay for a considerable time and till it retired on the approach of the reserve the charge successfully accomplished its object. It saved the position and probably the battlefield. The loss of the eight companies in the charge was 215 killed and wounded, more than 83 percent. 47 men were still in line and no missing. In self-sacrificing desperate valor this charge has no parallel in any war. The next day the regiment participated in repelling Pickett's charge losing 17 more men killed and wounded."

Marker: National Cemetery. Dedicated to the slain. Map reference **II I-5**

2nd NEW HAMPSHIRE INFANTRY
3rd Corps, 2nd Div., 3rd Brig.

Raised: Counties of Rockingham, Cheshire, Hillsborough, Merrimack and Strafford.
Organized: Camp Constitution, Portsmouth, N.H. M.I. June 10, 1861.
Commander: Col. Edward Lyon Bailey. b. Manchester, N.H., Dec. 10, 1841. Clerk in Manchester Post Office. Capt., Co. I, 2nd N.H. June 7, 1861. Wd. May 5, 1862 and slightly July 2, 1863. M.O. June 29, 1864. 2 Lt., 4th U.S. Infantry, Mar. 7, 1867. Dismissed by order of court-martial with rank of Capt., Oct. 15, 1893. d. Manchester, N.H., Mar. 12, 1930.
Number: 354　**Loss:** 20 k., 137 wd., 36 m.
Monument: Peach Orchard. Map reference **IV M-1**
"July 2, 1863."

5th NEW HAMPSHIRE INFANTRY
2nd Corps, 1st Div., 1st Brig.

Raised: Counties of Merrimack, Sullivan, Grafton, Carroll and Coos.
Organized: Camp Jackson, Concord, N.H. M.I. Oct. 1861.
Commander: Lt. Col. Charles Edward Hapgood. b. Shrewsbury, Mass., Dec. 11, 1830. Merchant in Amherst, N.H. Capt., Co. I, 5th N.H., Oct. 12, 1861. Wd. June 16, 1864. M.O. Col., Oct. 14, 1864. d. soldier's home in Chelsea, Mass., Sept. 24, 1909.
Number: 182　**Loss:** 27 k., 53 wd.
Monument: Ayres Ave. Map reference **V C-7**
"Here July 2, 1863 from 5 p.m. til 7 the 5th N.H. Vols. stood and fought. On this spot fell mortally wounded Edward E. Cross, Col. 5th N.H. Vols."
Marker: Sickles Ave. Map reference **V D-8**
"Left of Cross' Brigade."

12th NEW HAMPSHIRE INFANTRY
3rd Corps, 2nd Div., 1st Brig.

Raised: Counties of Belknap, Carroll, Grafton and Merrimack.
Organized: Camp Belknap, Concord, N.H. M.I. Aug.-Sept., 1862.
Commander: Capt. John F. Langley. b. Nottingham, N.H., Aug. 21, 1829. Machinist in Manchester, N.H. Capt., Co. F, 12th N.H., Sept. 5, 1862. M.O. Maj., Aug. 31, 1864. d. Amherst, N.H., Sept. 10, 1917.
Number: 245　**Loss:** 20 k., 70 wd., 2 m.
Monument: Emmitsburg Road. Map reference **IV F-4**
"It marched to this field on the night of the 1st. Fought here on the 2nd and supported the centre against Pickett's charge on the 3rd."

1st NEW HAMPSHIRE ARTILLERY, BATTERY A
Artillery Reserve, 3rd Volunteer Brigade

Raised: Hillsborough County.

Organized: Fair grounds, Manchester, N.H. M.I. Sept. 26, 1861.

Commander: Capt. Frederick Mason Edgell. b. Lyme, N.H., July 6, 1828. Grad. Kimball Union Academy, 1845. Mexican War service. Tinsmith in Manchester. Lt., Battery A, Sept. 26, 1861. Ended service as Maj., 1st N.H. Heavy Artillery, June 15, 1865. d. Georgetown, Mass., June 21, 1877.

Number: 4 Ordnance Rifles. 111 men.   **Loss:** 3 wd.

Monument: National Cemetery. Map reference **II H-6**

"On this ground Edgell's 1st New Hampshire Battery, Light Artillery, fired three hundred and fifty-three rounds of ammunition July 2nd and 3rd, 1863."

4th Michigan Infantry, DeTrobriand Avenue.

1st NEW JERSEY INFANTRY
6th Corps, 1st Div., 1st Brig.

Raised: Counties of Middlesex, Mercer, Warren, Camden, Union and Hudson.
Organized: Camp Olden, Trenton, N.J. M.I. May 21, 1861.
Commander: Lt. Col. William Henry, Jr. b. Stroudsburg, Penn., Dec. 15, 1836. Clerk, Oxford Furnace, N.J. Adjt., 1st N.J. Inf., June 6, 1861. Wd. Aug. 27, 1862. M.O. June 23, 1864. d. Fort Worth, Texas, Mar. 16, 1889.
Number: 292 **Loss:** None
Monument: No unit monument.
Marker: Near Brigade Monument. Map reference **IV J-15**

2nd NEW JERSEY INFANTRY
6th Corps, 1st Div., 1st Brig.

Raised: Counties of Essex, Passaic, Union and Sussex.
Organized: Camp Olden, Trenton, N.J. M.I. May 27, 1861.
Commander: Lt. Col. Charles Wiebecke. b. Gebstaedt, Prussia, Nov. 8, 1827. Served as pvt. in the Prussian army. Came to U.S. in 1851. Barber in Newark, N.J. Capt., Co. E, 2nd N.J., May 28, 1861. Killed in fighting at Myers Hill, Spotsylvania battlefield, Va., May 14, 1864. Buried for a time in the Fredericksburg National Cemetery.
Number: 405 **Loss:** 6 wd.
Monument: No unit monument. Regiment honored on New Jersey Brigade Monument.
Marker: Near Brigade Monument. Map reference **IV K-14**

3rd NEW JERSEY INFANTRY
6th Corps, 1st Div., 1st Brig.

Raised: Counties of Union, Gloucester, Burlington, Somerset, Cumberland, Camden, and Sussex. Also a large number from Philadelphia, Penn.
Organized: Camp Olden, Trenton, N.J. M.I. June 4, 1861.
Commander: Col. Henry Willis Brown (Barnes). b. Barnes in Boston, Mass. At age 22, he deserted his wife, changed name to Brown, and joined the 4th U.S. Artillery (1839-1844). Then he lived in Detroit, Mich., Philadelphia, Penn. and the gold fields of California. Clerk in Philadelphia at start of the war. Capt., Co. A, 3rd N.J., May 22, 1861. Wd. May 3, 1863 and during Spotsylvania, Va. Campaign. M.O. June 23, 1864. d. Deer Island, Mass., Oct. 25, 1892, age 76 years, 3 months, 20 days.
Number: 325 **Loss:** 2 wd.
Monument: No unit monument. Regiment honored on New Jersey Brigade Monument.
Marker: Near Brigade Monument. Map reference **IV L-14**

4th NEW JERSEY INFANTRY
Companies ACH, 6th Corps, 1st Div., Provost Guard
Companies BDEFGIK, Reserve Artillery Train Guard

Raised: Counties of Camden, Salem, Burlington and Union.
Organized: Camp Olden, Trenton, N.J. M.I. Aug. 19, 1861.
Commander: Maj. Charles Ewing. b. Trenton, N.J., June 6, 1841. Grandson of Chief Justice of New Jersey for whom he was named. Ship's officer on ocean going vessels. Ensign, 3rd N.J. Inf. (3 mos.). Capt., Co. B, 6th N.J Inf., Aug. 24, 1861. Transferred to 4th N.J., Feb. 20, 1863. Wd., May 3, 1863 and May 12, 1864. M.O. Mar. 17, 1865. d. Trenton, Mar. 14, 1872.
Number: 386 **Loss:** None
Monument: No unit monument. Regiment honored on New Jersey Brigade Monument.
Marker: School House Road. Location of train guard. Map reference III G-19

5th NEW JERSEY INFANTRY
3rd Corps, 2nd Div., 3rd Brig.

Raised: Counties of Hudson, Salem, Essex, Burlington, Monmouth and Passaic.
Organized: Camp Olden, Trenton, N.J. M.I. Aug. 22, 1861.
Commanders: Col. William Joyce Sewell. b. Castlebar, Ireland, Dec. 6 1835. Came to U.S. in 1851. Went to sea for two voyages to China and Australia. Merchant in Chicago, Ill. 1860 moved to Camden, N.J. Capt., Co. C, 5th N.J., Aug. 28, 1861. Wd. May 3, 1863 and July 2, 1863. M.O. July 6 1864. Served as Col., 38th N.J. until M.O. June 30,1865. Awarded Medal of Honor for actions May 3, 1863 at Chancellorsville, Va. Member of State Senate. Elected to U.S. Senate from N.J. d. Camden, Dec. 27, 1901. When Sewell wd., Capt. Thomas C. Godfrey took command. b. Philadelphia, Penn., February 6, 1835. Painter in Allowaystown, N.J. 2 Lt., Co. F, 5th N.J., Aug. 22, 1861. M.O. Sept. 7, 1864. Served with Veteran Volunteers unit until discharged Mar. 13, 1866. d. Allowaystown, N.J., June 13, 1867. Capt. Henry Harrison Woolsey took command from Godfrey on July 3, 1863. b. Pennington, N.J., April 1, 1837. Grad. Princeton College, N.J., 1856. Lawyer in Trenton. 2 Lt., Co. E, 5th N.J., Aug. 19, 1861. Wd. Aug. 29, 1862, slightly July 2, 1863 and June 18, 1864 near Petersburg, Va. The final wound resulted in his death on June 19, 1864. Last words: "I die in a glorious cause...."
Number: 221 **Loss:** 13 k., 65 wd., 16 m.
Monument: Emmitsburg Road. Map reference IV D-4
"The regiment first held the skirmish line 400 yards to the front and left of this spot, and afterwards took position in the line of battle here."

6th NEW JERSEY INFANTRY
3rd Corps, 2nd Div., 3rd Brig.

Raised: Counties of Camden, Hudson, Hunterdon, Mercer and Burlington.

Organized: Camp Olden, Trenton, N.J. M.I. Aug. 19, 1861.

Commander: Lt. Col. Stephen Rose Gilkyson. b. probably Makefield, Penn., 1833 or 1834. Lumber merchant in Hightstown, N.J. Capt., Co. A, 6th N.J., Aug. 19, 1861. Wd. Aug. 29, 1862 and May 6, 1864. M.O. Sept. 7, 1864. Subsequent service as Col. in 40th N.J. Inf. until M.O. July 13,1865. d. Jan. 31, 1892.

Number: 246   **Loss:** 1 k., 32 wd., 8 m.

Monument: Crawford Ave. Map reference **V E-8**

"Engaged here July 2,1863, being detached from the brigade. Supported batteries on Cemetery Ridge July 3."

7th NEW JERSEY INFANTRY
3rd Corps, 2nd Div., 3rd Brig.

Raised: Counties of Essex, Morris, Passaic, Hudson, Sussex and Warren.

Organized: Camp Olden, Trenton, N.J. M.I. Sept. 3, 1861.

Commanders: Col. Louis Raymond Francine. b. Philadelphia, Penn., Mar. 26, 1837. Attnd. military schools in Flushing, Long Island, New York and in France. Civil Engineer in Camden, N.J. Capt., Co. A, 7th N.J., Aug. 23, 1861. Wd. July 2, 1863. d. of wounds July 16, 1863. When Francine wounded, Maj. Frederick Cooper took command of regiment. b. Ireland. Ferryman in Jersey City, N.J. Ensign, 2nd N.J. Inf. (3 mos.). Capt., Co. F, 7th ; N.J., Sept. 2, 1861. Wd. June 19,1864. M.O. Maj., Sept. 30, 1864. d. Feb. 14, 1874, age 42 years, 1 month, 8 days.

Number: 331   **Loss:** 15 k., 86 wd., 13 m.

Monument: Sickles Ave. Map reference **IV K-3**

"First position 300 yards N.E. of this. Heavily engaged there. Moved here to reinforce Graham's Brigade. Here Colonel Francine fell."

8th NEW JERSEY INFANTRY
3rd Corps, 2nd Div., 3rd Brig.

Raised: Counties of Essex, Hudson and Hunterdon.

Organized: Camp Olden, Trenton, N.J. M.I. Sept. 14, 1861.

Commanders: Col. John Ramsey. b. N.Y. City, Oct. 7, 1838. Cigar maker in Jersey City, N.J. Lt., 2nd N.J. Inf. (3 mos.) Capt., 5th N.J. Inf. Col., 8th N.J., April 6, 1863. Wd. July 2, 1863 and June 16, 1864. M.O. July 17, 1865. d. Jersey City, Feb. 11, 1901. Buried Arlington National Cemetery. When Ramsey wd., Capt. John G. Langston took command of regiment. b. England around 1828. Carpenter in Jersey City, N.J. Capt., Co. K, 8th N.J., Sept. 27, 1861. Wd. slightly May 5, 1862. Dismissed by order of court martial. Allowed to resign and was M.O. Sept. 21, 1864. d. Philadelphia, Penn., Feb. 10, 1903.

Number: 148   **Loss:** 7 k., 38 wd., 2 m.

Monument: De Trobriand Ave. Map reference **V C-5**

"Engaged here July 2,1863, being detached from the Brigade. Supported batteries on Cemetery Ridge July 3."

11th NEW JERSEY INFANTRY
3rd Corps, 2nd Div., 1st Brig.

Raised: Counties of Mercer, Morris, Union, Passaic, Essex, Middlesex and Hudson.

Organized: Camp Perrine, Trenton, N.J. M.I. Aug. 15,1862.

Commanders: Col. Robert McAllister. b. Lost Creek Valley, Juniata County, Penn., June 1, 1813. Militia officer. Railroad construction business in Oxford Furnace, N.J. Lt. Col., 1st N.J. Inf. Col., 11th N.J., Aug. 18, 1862. Wd. July 2, 1863. M.O. June 6, 1865. d. Belvidere, N.J., Feb. 23, 1891. When McAllister wounded, Capt. Luther Martin took command. b. Penn., 1826. Printer in Elizabeth, N.J. Sergt., 1st N.J. Inf. Capt., Co. D, 11th N.J., Aug. 6, 1862. Soon after McAllister wounded, Martin was hit in the foot. Trying to go to the rear, he was struck in the thigh. Struggling, he then was hit in the chest, resulting in his death. Capt. William H. Loyd briefly took command from Martin. b. Philadelphia, Penn., Jan. 27, 1839. Stock broker in Philadelphia. 3 mos. service. 2 Lt., Co. I, 11th N.J., Aug. 7, 1862. Wd. July 2, 1863 and Oct. 27, 1864. M.O. Jan. 13, 1865. Lived in Ardmore, Penn. d. June 2, 1907. After Loyd out of action, Lt. and Adjt. John Schoonover assumed command. b. Bushkill, Penn., Aug. 12, 1839. Teacher in Belvidere, N.J. Pvt., 1st N.J. Inf. In Aug. 1862 he was transferred to 11th N.J. Wd. July 2, 1863 and June 8, 1864. M.O. Lt. Col., June 6, 1865. d. April 13, 1930. Late in the fight on July 2, Capt. Samuel Tucker Sleeper commanded regiment. b. Pemberton, N.J., Mar. 15, 1823. Tailor in Shrewsbury, N.J. Lt., Co. I, 11th N.J., Aug. 6, 1862. Killed in Battle of Spotsylvania Court House Va., May 12, 1864.

Number: 344 (monument has 275 engaged).   **Loss:** 17 k., 124 wd., 12 m.

Monument: Emmitsburg Road. Map reference **IV F-3**

"This stone marks the spot reached by the right of the regiment, the left extending towards the southeast. The position was held under a severe fire, which killed or disabled nearly three-fifths of the regiment, including every officer present above the rank of lieutenant."

Excerpt from Schoonover's monument dedication speech, 1888 —

"Near the spot upon which we now stand the regiment took up its position on that eventful afternoon of July 2nd 1863. About 3 o'clock the enemy opened with artillery. For an hour or more the earth trembled with the jar of guns, and tons of metal were hurled over and fell around us. At this time our right was resting just in rear of the Smith house, over there, the line extending down through the [Smith] orchard, the left being under the brow of the hill. Major Kearny, who was standing near me on the left of the regiment excitedly exclaimed: 'we are going to have a fight.' A moment later he was mortally wounded and carried to the rear. We were now receiving a fire from Wilcox's brigade from beyond the Emmitsburg Pike, and from Barksdale's brigade, which was advancing in full view just down by the road which leads to Little Round Top."

12th NEW JERSEY INFANTRY
2nd Corps, 3rd Div., 2nd Brig.

Raised: Counties of Salem, Burlington, Camden, Gloucester and Cumberland.
Organized: Camp Stockton, Woodbury, N.J. M.I. Sept. 4, 1862.
Commander: Maj. John T. Hill. b. New Brunswick, N.J., July 1836. Clerk in Park Bank of N.Y. City. Capt., 11th N.J. Inf. Maj., 12th N.J., April 23, 1863. M.O. Feb. 24, 1864. Lived in New Brunswick. d. Mar. 1, 1891.
Number: 532   **Loss:** 23 k., 83 wd., 9 m.
Monument: Hancock Ave. Map reference II F-9
Marker: Bliss Farm site. Map reference II A-8
"Erected by the State of New Jersey, 1888, in honor of the 12th Regiment of volunteers, a detachment of which in the afternoon of July 2, 1863, charged the Bliss house and barn here capturing the enemy's skirmish reserve of 7 officers and 85 men stationed therein. On the morning of July 3 another detachment of the regiment charged, capturing the buildings, one officer and one man, and driving back the skirmish reserve. The regiment lost in their charges 60 officers and men."

13th NEW JERSEY INFANTRY
12th Corps, 1st Div., 3rd Brig.

Raised: Counties of Essex, Hudson, and Passaic.
Organized: Camp Frelinghuysen, Newark, N.J. M.I. Aug. 25, 1862.
Commander: Col. Ezra Ayers Carman. b. near Oak Tree, Middlesex County, N.J., Feb. 27, 1834. Grad. Western Military Institute in Kentucky, 1855. Taught there (in 1855, the school moved to Nashville, Tenn.). Accountant in Newark, N.J. at start of the war. Lt. Col., 7th N.J. Inf. Col., 13th N.J., July 8, 1862. Wd. May 5, 1862. M.O. June 8, 1865. d. Washington, D.C., Dec. 25, 1909. Buried in Arlington National Cemetery.
Number: 360   **Loss:** 1 k., 20 wd.
Monument: Carman Ave. Map reference III I-10
"Thirteenth Regiment N.J.V. reached this battle-field 5 p.m. July 1st 1863, and with the brigade went into position on the north side of Wolf Hill. During the night occupied a position in support of Battery M, First N.Y. Artillery. July 2, in morning held position near Culp's Hill; In afternoon marched to relief of Third Corps near Round Top; At night returned to right of the army. July 3, occupied position marked by this monument supporting Second Massachusetts and Twenty-seventh Indiana in their charge on Confederate flank. In the evening moved to extreme right to support Gregg's Cavalry."
Markers: Carman Ave. Position of companies G & I. Map reference III I-10

15th NEW JERSEY INFANTRY
6th Corps, 1st Div., 1st Brig.

Raised: Counties of Sussex, Warren, Hunterdon, Morris and Somerset.

Organized: Camp Fair Oaks, Flemington fair grounds, Flemington, N.J. M.I. Aug. 25, 1862.

Commander: Col. William Henry Penrose. b. Madison Barracks, Sacket's Harbor, N.Y., Mar. 10, 1832. Father was an officer in the Regular army. Attnd. Dickinson College, Penn. Civil engineer in Niles, Michigan. 2 Lt., 3rd U.S. Inf. Col., 15th N.J., April 18, 1863. Wd. Oct. 19, 1864. M.O. Brig. Gen., Jan. 15, 1866. Capt., 3rd U.S. Inf. Served in Regular army until retirement, Mar. 10, 1896. d. Salt Lake City, Utah, Aug. 29, 1903. Buried Arlington Nat'l. Cem.

Number: 441 **Loss:** 3 wd.

Monument: No unit monument. Regiment honored on the New Jersey Brigade Monument.

Marker: Near Brigade monument. Map reference **IV K-15**

1st NEW JERSEY ARTILLERY, BATTERY A
Artillery Reserve, 4th Volunteer Brigade

Other Names: "Hexamer's"

Raised: Hudson County.

Organized: Hoboken, N.J. M.I. Aug. 12, 1861.

Commander: Lt. Augustin N. Parsons. b. Granville, Mass., around 1830. Enrolled in Regular service at Cincinnati, Oh. Occupation listed as carpenter. Service in 2nd U.S. Artillery. Lt., Battery A, 1st N.J., July 28, 1862. M.O. Capt., June 22, 1865. In 1891 he lived in Summit, Tyler County, Texas.

Number: 6 10-pdr. Parrotts. 116 men. **Loss:** 2 k., 7 wd.

Monument: Hancock Ave. Map reference **IV D-13**

"Battery A, 1st New Jersey Artillery, from its position in reserve S.W. of Powers Hill galloped into action at 3 p.m., July 3, 1863. Fired 120 rounds shrapnel at Pickett's column, and then 80 rounds shell at a battery in left front. Position in action, 45 yards E. of this stone."

1st NEW JERSEY ARTILLERY, BATTERY B
3rd Corps Artillery Brigade

Raised: Essex County. Organized Camp Olden, Trenton, N.J. M.I. Sept. 3, 1861.

Commanders: Capt. Adoniram Judson Clark. b. Fayetteville, N.Y., Oct. 8, 1838. Carpenter and student of medicine in Newark, N.J. Sergt., 1st N.J. (3 mos.). Lt., Battery B, 1st N.J., April 27, 1863. M.O. June 16, 1865. Held various civil posts in Newark including Chief of Police. d. Newark, July 24, 1913.

Number: 6 10-pdr. Parrotts. 143 men. **Loss:** 1 k., 16 wd., 3 m.

Monument: Sickles Ave. Map reference **IV K-3**

"Fought here from 2 until 7 O'Clock on July 2,1863, firing 1,300 rounds of ammunition."

Marker: Tablet, Hancock Ave. Map reference **IV 1-13**

"Position July 3, 1863."

1st NEW JERSEY CAVALRY
Cavalry Corps, 2nd Div., 1st Brig.
Companies E and G in defenses of Washington, D.C.
Company L, 6th Corps Headquarters

Other Names: "Sixteenth"

Raised: Counties of Mercer, Sussex, Burlington, Hudson, Essex and Middlesex.

Organized: Trenton, N.J. M.I. Sept. 1861.

Commander: Maj. Myron Holley Beaumont. b. Wayne County, N.Y., around 1837. Father Abram was a prominent doctor in Lyons, Wayne County. Entered U.S. cavalry service in 1856, occupation at enlistment, printer. Start of war resided in Rahway, N.J. Lt., 3rd N.J. Infantry (3 mos.). Maj., 1st N.J. Cavalry, Sept. 16, 1861. Wd. Feb. 6, 1865. M.O. Col., July 24, 1865. After the war he deserted his family and headed west with his brother in law's wife. He was married at least once more without the benefit of legal divorce. After years of criminal business activities, he ended up in Ukiah, California going by the name of Thomas B. Edwards. There he committed suicide on Feb. 6, 1878.

Number: 269 **Loss:** 9 wd.

Monument: Gregg Ave. East Cavalry Battlefield. No map reference.

"Fought here July 3, 1863, both mounted and dismounted, holding this position several hours. Assisted in repelling the charges of the enemy's cavalry."

Marker: Sedgwick Ave. Indicates position of Company L. Not found.

110th and 62nd Pennsylvania Infantry, DeTrobriand Avenue.

8th NEW YORK INFANTRY
Independent Company
11th Corps Headquarters

Other Names: "First German Rifles"
Raised: New York City
Organized: Palace Garden, N.Y. City. M.I. April 23, 1861 for two years. Reorganized May 1863 into one company.
Commander: Lt. Hermann Foerster. Pvt., Co. K, 8th N.Y., April 23, 1861. Age 30. Resigned Mar. 25, 1864. Father lived in Europe.
Number: 40   **Loss:** None
Monument: None

10th NEW YORK BATTALION
4 Companies, ABCD
2nd Corps, 3rd Div., Provost Guard

Other Names: "National Zouaves"
Raised: New York City
Organized: Sandy Hook, New Jersey. M.I. April 27 to May 7, 1861 for two years. Reorganized April 26, 1863 into four companies.
Commander: Maj. George Faulkner Hopper. b. N.Y. City, April 26, 1824. Paperhanger and volunteer fireman in N.Y. City. Capt., Co. H, 10th N.Y. April 30, 1861. M.O. Lt. Col., June 30, 1865. d. Paskack, N.J., Aug. 4, 1891.
Number: 98 **Loss:** 2 k., 4 wd.
Monument: Map reference II G-10
"Held this position with 8 officers and 90 enlisted men as provost guard Hays's Division during Pickett's charge July 3, 1863."

12th NEW YORK INFANTRY
2 Companies, DE
5th Corps Headquarters

Raised: New York City
Organized: From consolidation of 12th Militia and 12th N.Y. Volunteers in Feb. 1862. Reorganized into two companies May 17, 1863.
Commander: Capt. Henry Wines Ryder. b. N.Y. City, Nov. 30, 1833. Militia service. Capt., Co. E, 12th N.Y., Jan. 29, 1862. Wd. Aug. 30, 1862. M.O. Maj., Aug. 21, 1865. Late in life was a merchant in Newark, N.J. where he died Dec. 1, 1910.
Number: 117   **Loss:** None
Monument: Little Round Top, 12th N.Y. and 44th N.Y. combined.
Map reference **V F-11**

29th NEW YORK INFANTRY
Independent Company
11th Corps, 2nd Div., Provost Guard

Other Names: "First German Infantry." "Astor Rifles."
Raised: New York City
Organized: Conrad's Elm Park, N.Y. City. M.I. June 4, 1861 for two years. Reorganized into one company June 1, 1863.
Commander: 2 Lt. Hans Von Brandis. b. Hanover, Germany, July 22, 1834. German officer and insurance salesman. 2 Lt., Co. A, 29th N.Y., Dec. 23, 1862. M.O. Lt., May 17, 1864. d. Brooklyn, N.Y., Dec. 16, 1913. Buried Cypress Hills National Cemetery.
Number: 36 **Loss:** 2 wd., 4 m.
Monument: None

33rd NEW YORK INFANTRY
Detachment
6th Corps, 2nd Div., 3rd Brig.

Raised: Men at the Battle of Gettysburg from Monroe County.
Organized: Elmira, N.Y. M.I. to date May 22, 1861 for two years. In May 1863 remaining members attached to 49th N.Y. Not transferred until Oct. 1, 1863.
Commander: Capt. Henry Judson Gifford. b. New York, Sept. 8, 1836. "Agent", Bergen, N.Y. 2 Lt., 13th N.Y. Lt., Co. D, 33rd N.Y., Aug. 30, 1861. Wd. Sept. 17, 1862. M.O. July 3, 1865. d. Norfolk, Va., Dec. 29, 1909. Buried Arlington National Cemetery.
Number: 69 **Loss:** None
Monument: None

39th NEW YORK INFANTRY
4 Companies, ABCD
2nd Corps, 3rd Div., 3rd Brig.

Other Names: "Garibaldi Guards"
Raised: New York City
Organized: New York City. M.I. to date May 28, 1861.
Commander: Maj. Hugo Hillebrandt. b. Hungary, 1832. Joined revolutionary army under Louis Kossuth. Fled to U.S. Returned to fight with Garibaldi in Italy. Returned to U.S. in 1860. Civil Engineer. Lt. and Adjt., 39th N.Y., June 6, 1861. Wd. July 3, 1863. M.O. Dec. 10, 1863. Served in Veteran Reserve Corps. Entered U.S. foreign service. d. Brooklyn, N.Y., April 4, 1896.
Number: 322 **Loss:** 15 k., 80 wd.
Monument: Hancock Ave. Map reference II F-10
  "This regiment at about 7 o'clock p.m. July 2, 1863 being ordered to support General Sickles' line, charged and drove the enemy recapturing the guns

and equipment of Battery I, 5th U.S. artillery. A stone tablet marks the place where this incident occurred."

Marker: Stone tablet, U.S. Ave. Map reference **IV J-11**

40th NEW YORK INFANTRY
3rd Corps, 1st Div., 3rd Brig.

Other Names: "Mozart Regiment"

Raised: Men at Gettysburg predominantly from New York City and Onondaga County.

Organized: Camp Wood, Yonkers, N.Y. M.I. June 27, 1861. Formed under the auspices of the Mozart Hall Committee, a New York City political faction.

Commander: Col. Thomas Washington Egan. b. N.Y. City, 1834 or 1836. Clerk in N.Y. City. Lt. Col., 40th N.Y., July 1, 1861. Wd. slightly at Gettysburg and June 18,1864. M.O. Brig. Gen., Jan. 15, 1866. After war worked in New York Customs house. d. N.Y. City, Feb. 24, 1887. Buried Cypress Hills National Cemetery.

Number: 606 **Loss:** 23 k., 120 wd., 7 m.

Monument: Crawford Ave. Map reference **V F-9**
"July 2, 1863; 4:30 p.m."

41st NEW YORK INFANTRY
9 Companies
11th Corps, 1st Div., 1st Brig.

Other Names: "De Kalb Regiment"

Raised: New York City, Philadelphia, Penn., and Essex County, N.J.

Organized: Conrad's Park, N.Y. City. M.I. June 9,1861. Company F became 9th N.Y. Independent battery early in the war.

Commander: Lt. Col. Heinrich Detleo Von Einsiedel. German military officer from the Dresden area. Arrived in U.S. around 1858. Capt., Co. E, 41st N.Y., June 6, 1861, age 33. d. at Petersburg, Va. of typhoid fever Aug. 23, 1865. Buried Poplar Grove National Cemetery.

Number: 218 **Loss:** 15 k., 58 wd., 2 m.

Monument: Wainwright Ave. Map reference **II L-4**
"July 2, 3, 1863."

42nd NEW YORK INFANTRY
2nd Corps, 2nd Div., 3rd Brig.

Other Names: "Tammany Regiment"

Raised: New York City

Organized: Camp Tammany, Great Neck, Long Island, N.Y. M.I. June 22-28, 1861. Formed under the auspices of the Tammany Society and the Union Defense League of N.Y. City.

Commander: Col. James Edward Mallon. b. Brooklyn, N.Y., Sept. 12, 1836. Wholesale commission business in N.Y. City. Belonged to 7th N.Y. State Militia. 2 Lt., 40th N.Y. Wd. May 31, 1862. Maj., 42nd N.Y., Aug. 2, 1862. K. Oct. 14, 1863 at Bristoe Station, Va.

Number: 197   **Loss:** 15 k., 55 wd., 4 m.

Monument: Hancock Ave. Map reference **II F-13**

On monument is the figure of Tammany, the Delaware Indian chief who sided with the Americans during the American Revolution.

"July 2,1863. Went to support of 3rd Corps, about 5 p.m. Held this position July 3, and assisted in repulsing the assault of Pickett's Division."

43rd NEW YORK INFANTRY
6th Corps, 2nd Div., 3rd Brig.

Other Names: "Vinton Rifles"

Raised: New York City and Counties of Albany, Montgomery and Otsego.

Organized: Industrial School Barracks, Albany, N.Y. M.I. Sept. 22, 1861.

Commander: Lt. Col. John Wilson. b. Albany, N.Y., Dec. 29, 1838. Florist and nurseryman in Albany, N.Y. Capt., Co. A, 43rd N.Y., Aug. 25, 1861. Mort. wd. May 6, 1864 in the Battle of the Wilderness, Va. d. May 7,1864.

Number: 403   **Loss:** 2 k., 2 wd., 1 m.

Monument: Neill Ave. Map reference **III N-16**

"Arrived on field 4 p.m., July 2,1863. Held this position from the morning of July 3rd until close of battle."

44th NEW YORK INFANTRY
5th Corps, 1st Div., 3rd Brig.

Other Names: "People's Ellsworth Regiment"

Raised: Counties of Erie, Albany and Oneida. Representatives from almost every county in the state.

Organized: Industrial School Barracks, Albany, N.Y. M.I. Aug.-Sept., 1861. Formed by the Ellsworth Association of the State of New York.

Commanders: Col. James Clay Rice. b. Worthington, Mass., Dec. 27, 1829. Grad. Yale College, 1854. Lawyer, N.Y. City. Lt., 39th N.Y. Lt. Col., 44th N.Y., Sept. 13, 1861. Killed May 10, 1864 at Spotsylvania Court House, Va. Held the rank of Brig. Gen. at death. Last words-"turn me over that I may die with my face to the enemy." When Rice took brigade, Lt. Col. Freeman Conner took command of regiment. b. Exeter, New Hampshire, Mar. 2, 1836. Commission merchant in Chicago, Ill. Member of Ellsworth's United States Zouave Cadets. Lt., 11th N.Y. Capt., Co. D, 44th N.Y., Sept. 11, 1861. Wd. Dec. 13, 1862 and May 8, 1864. M.O. Oct. 11, 1864. Died on a city street in Chicago, Mar. 28, 1906.

Number: 460   **Loss:** 26 k., 82 wd., 3 m.

Monument: Little Round Top. 12th N.Y. and 44th N.Y. monuments combined. Map reference **V F-11**

"The 44th N.Y. Infantry, Lieut. Colonel Freeman Conner commanding, held position about 100 feet in advance of this monument, designated by a marker, from about 5 p.m. July 2, to about 11 a.m. July 3, 1863. At noon of July 3rd, was placed in reserve at the right of Little Round Top where it remained until the close of the battle."

Marker: Not found

45th NEW YORK INFANTRY
11th Corps, 3rd Div., 1st Brig.

Other Names: "Fifth German Rifles." "Howe's Rifles."
Raised: New York City
Organized: Landmann's Park, N.Y. City. M.I. Sept. 9, 1861.
Commanders: Col. George Karl Heinrich Wilhelm Von Amsberg. b. Hildesheim, Germany, June 24, 1821. After commanding troops in Hungarian revolution, he fled to U.S. in the late 1850s. Riding master in Hoboken, N.J. Joined 5th N.Y. State Militia. Maj. 45th N.Y., May 1, 1861. M.O. Jan. 22, 1864. Hotel business in Hoboken, where he died Nov. 21, 1876. When Von Amsberg took command of brig., Lt. Col. Adolphus Dobke led regiment. b. Germany. Police officer in N.Y. City. He wrote that he was also a merchant before the war. Capt., Co. D, 45th N.Y., Sept. 20, 1861. Wd. Second Bull Run. M.O. Oct. 15, 1865. d. Jersey City, N.J., Mar. 1904, age 82 years, 11 months, 11 days.
Number: 447 **Loss:** 11 k., 35 wd., 178 m.
Monument: Howard Ave. Map reference I E-10
"This regiment went into action about 11:30 a.m., July 1st 1863 by deploying four companies as skirmishers under Captain Irsch, about one hundred yards to the rear of this monument, they advanced supported by the other six companies under Lt. Col. Dobke, about five hundred and forty yards under a terrific artillery and sharpshooters fire to a point indicated by marker in front. The regiment also assisted in repelling a charge on the flank of the 1st Corps to the left, capturing many prisoners. Covered retrograde movement into town, fighting through the streets, where Major Koch fell desperately wounded. A portion of the regiment was cut off and took shelter in connecting houses and yards on Chambersburg Street west of the town square, holding the enemy at bay, until about 5:30 p.m. when they surrendered, after having destroyed their arms and accoutrements. On July 2, the remnant of the regiment was exposed to a heavy artillery fire on Cemetery Hill, and in the evening moved hastily to Culp's Hill and assisted in repulsing an attack on Greene's Brigade, 12th Corps (see markers on Culp's and Cemetery Hills). On 3d it was again exposed to artillery and sharpshooter's fire, whereupon Sergt. Link, with volunteers, dislodged the enemy's sharpshooters in the edge of the town, nearly all the small attacking party being killed or wounded in the effort."

Markers: Only one marker found. McClean Farm Lane. Map reference I C-8. McClean Farm Lane stone marks advance position of regiment.

49th NEW YORK INFANTRY
6th Corps, 2nd Div., 3rd Brig.

Other Names: "Second Buffalo"
Raised: Counties of Chautauqua, Erie, Westchester and Niagara.
Organized: Fort Porter, Buffalo, N.Y. M.I. Sept. 18, 1861.
Commander: Col. Daniel Davidson Bidwell. b. Black Rock, N.Y., Aug. 12, 1819. Lawyer and police justice in Buffalo. Militia officer. Col., 49th N.Y., Oct. 21, 1861. Mort. wd., Brig. Gen., Oct. 19, 1864 at Battle of Cedar Creek, Va. Last words, "I have tried to do my duty."
Number: 414 **Loss:** 2 wd.
Monument: Neill Ave. Map reference **III M-16**
"Held this position July 3, 1863."

52nd NEW YORK INFANTRY
2nd Corps, 1st Div., 3rd Brig.

Other Names: "German Rangers." "Sigel Rifles."
Raised: New York City
Organized: Camp Washington at Quarantine Grounds, Staten Island, N.Y. M.I. Oct. 25, 1861.
Commanders: Lt. Col. Charles Godfrey Freudenberg. b. Baden, Germany, May 1, 1833. At age 15 joined revolutionary movement in Germany. Came to U.S. a few years before the Civil War. Capt., Co. A, 52nd N.Y., Aug. 3, 1861. Wd. Battle of Fair Oaks and July 2, 1863. M.O. Maj., Veteran Reserve Corps, Mar. 9, 1866. Joined 45th U.S. Infantry. Retd. 1870. d. Portland Flats, Wash., D.C., Aug. 28, 1885. Maj. Edward Venuti took command from Freudenberg at Gettysburg. b. Italy 1825. Capt. 39th N.Y. May. 52nd N.Y., Feb. 9, 1863. Wd. June 8, 1862. Killed July 2, 1863. Command fell to Capt. William Scherrer. b. 1834 or 1835 in Germany. Lived in N.Y. City at start of war. 2 Lt., Co. B, 52nd N.Y., Nov. 1, 1861. Wd. Oct. 14, 1863. d. May 26, 1864 of wounds received May 12, 1864 near Spotsylvania Court House, Va.
Number: 134 **Loss:** 2 k., 26 wd., 10 m.
Monument: Sickles Ave. Map reference **V B-6**
"July 2nd 1863, 6 to 7 p.m."

54th NEW YORK INFANTRY
11th Corps, 1st Div., 1st Brig.

Other Names: "Hiram Barney Rifles." "Black Rifles."
Raised: New York City
Organized: U.S. arsenal, Hudson City, N.J. M.I. Sept. 5 - Oct. 16, 1861.
Commander: Maj. Stephen Kovacs. b. Hungary. Joined Hungarian revolution serving on staff of Louis Kossuth. Fled with Kossuth to U.S. in 1851. Clerk in N.Y. City. Capt., Co. K, 54th N.Y., Oct. 6, 1861. Captured July 1, 1863. M.O. June 20, 1865. d. April 15, 1884 in N.Y. City, age 59 years, 9 months, 5 days.

Lt. Ernst Both took command from Kovacs at Gettysburg. b. Holstein, Germany, 1830 or 1831. Came to U.S. in 1859. Clerk in N.Y. City. Served in 5th N.Y. State Militia. Pvt., Co. C, 54th N.Y., Sept. 17, 1861. M.O. Capt., April 14, 1866. d. soldiers home in Elizabeth City, Va., Mar. 14, 1888.

Number: 216 **Loss:** 7 k., 47 wd., 48 m.

Monument: Wainwright Ave. Map reference **II L-3**

Figure on stone represents 20 yr. old shoemaker Heinrich Michel, who d. July 2, 1863, carrying the colors. "July 1st skirmishing on extreme right near Rock Creek. July 2nd at sunset, severe fighting in this position. July 3rd held same position."

Marker: Near Rock Creek. Map reference **I B-16**

57th NEW YORK INFANTRY
2nd Corps, 1st Div., 3rd Brig.

Other Names: "National Guard Rifles." "Clinton Rifles." "Zook's Voltigeurs."

Raised: New York City. Counties of Dutchess and Oneida.

Organized: Camp Lafayette, New Dorp, Staten Island, N.Y. M.I. Aug. 12-Nov. 19, 1861.

Commander: Lt. Col. Alford B. Chapman. b. N.Y. City, Aug. 1, 1835. Merchant of "fancy goods" in N.Y. City. Militia service. Capt., Co. A, 57th N.Y., Oct. 19, 1861. Wd., Dec. 11, 1862. Killed May 5, 1864 in the Battle of Wilderness, Va. After being wounded in the Wilderness he wrote, "Dear Father; I am mortally wounded. Do not grieve for me. My dearest love to all - Alford."

Number: 179 **Loss:** 4 k., 28 wd., 2 m.

Monument: Sickles Ave. Map reference **V B-6**

"Engaged the enemy here July 2, 1863. July 3, on Cemetery Ridge resisting Pickett's attack."

58th NEW YORK INFANTRY
11th Corps, 3rd Div., 2nd Brig.

Raised: New York City

Organized: Turtle Bay Brewery at about 44th Street on East River, N.Y. City. M.I. Aug.-Nov. 1861.

Commanders: Lt. Col. August Otto. b. Holstein, Germany. Lt. Col., 58th N.Y., age 32, May 29, 1863. July 2, 1863 assigned to staff of Gen. Schurz. Resigned April 5, 1864. Capt. Emil Koenig assumed command from Otto. b. Prussia. In 1861 his age was 25 and his occupation was listed as soldier. Sergt., Co. l, 58th N.Y., Sept. 18, 1861. M.O. Oct. 1, 1865.

Number: 222 **Loss:** 2 k., 15 wd., 3 m.

Monument: Howard Ave. Map reference **I C-14**

"Two companies of the regiment held this position July 1, 1863 until ordered to Cemetery Hill. Were there joined by the other companies and engaged on the 2d and 3d. After the repulse of Pickett's Charge, skirmished into Gettysburg."

59th NEW YORK INFANTRY
4 Companies, ABCD
2nd Corps, 2nd Div., 3rd Brig.

Other Names: "Union Guards"

Raised: New York City, Counties of Richland, Oh. and Lewis, N.Y.

Organized: Camp Washington, Quarantine Grounds, Staten Island, N.Y. M.I. Aug. 2-Oct. 30, 1861.

Commanders: Lt. Col. Max A. Thoman. b. Germany. Spoke four languages. Seeking promotion he wrote, "I received a military education in the lyceum of Hanover, Germany from 1845-1848. I served as Lt. and Adjt. of the 9th Infantry and Jann's Corps in the 3 years war of Schleswig-Holstein against Danemark, 1848-1851 gaining the experience of 21 different engagements. I served in the English Foreign Legion in 1855 not actively in the field but assisted to form it. From 1856-1858 I resided in Central America during Walker's Expedition where I learned most valuable lessons about warfare in a southern or tropical climate!" Liquor salesman in N.Y. City before the war. Capt., Co. C, 59th N.Y., Oct. 8, 1861, age 31. Wd. Sept. 17, 1862 and mortally July 2, 1863. d. July 11, 1863. Quoted as saying, "Boys, bury me on the field." Buried Gettysburg Nat'l Cem. Capt. William McFadden. b. Mt. Vernon, Oh., Nov. 25, 1827. Merchant, Belleville, Oh. Pvt., Co. I, 59th N.Y., Sept. 21, 1861. M.O. Maj., Jan. 12, 1865. d. Clinton Twp., Knox Co., Oh., May 20, 1909.

Number: 182   **Loss:** 6 k., 28 wd.

Monument: Hancock Ave. Map reference II E-13

"Four companies of this regiment held this position July 2 and 3, 1863, where Max A. Thoman, Lieut. Colonel in command, fell mortally wounded."

60th NEW YORK INFANTRY
12th Corps, 2nd Div., 3rd Brig.

Other Names: "St. Lawrence Regiment"

Raised: Counties of St. Lawrence and Franklin.

Organized: Camp Wheeler, Ogdensburg, N.Y. M.I. Oct. 30, 1861.

Commander: Col. Abel Godard. b. De Kalb, N.Y., June 26, 1835. Law student in Richville, N.Y. Capt., Co. K, 60th N.Y., Oct. 19, 1861. Discharged for disability Sept. 13, 1863. Real estate agent in Richville. d. Richville, July 25, 1891.

Number: 273   **Loss:** 11 k., 41 wd.

Monument: Slocum Ave. Map reference III F-3

"July 2 and 3, 1863."

Marker: Location of Co. I. Lists all members of the company in the battle. Slocum Ave. Map reference III F-3

61st NEW YORK INFANTRY
2nd Corps, 1st Div., 1st Brig.

Other Names: "Clinton Guards"

Raised: New York City and Madison County.

Organized: Camp Harris, near Ft. Tompkins, Staten Island, N.Y. M.I. Sept.-Nov., 1861.

Commander: Lt. Col. Knut Oscar Broady. b. Upsala, Sweden, May 28, 1832. Joined Swedish artillery and later navy. Arrived U.S. in 1854. Start of war was a student at Madison University in Hamilton, N.Y. Capt., Co. C, 61st N.Y., Sept. 19, 1861. Wd. Aug. 25, 1864. M.O. Oct. 29, 1864. Missionary with American Baptist Society in Stockholm, Sweden. d. Stockholm, Mar. 13, 1922.

Number: 148   **Loss:** 6 k., 56 wd.

Monument: Wheatfield. Map reference **V B-8**

"This position held by the 61st Regt. N.Y. Infy. on the afternoon of July 2, 1863."

62nd NEW YORK INFANTRY
6th Corps, 3rd Div., 3rd Brig.

Other Names: "Anderson's Zouaves"

Raised: New York City

Organized: Camp Sumter, Union Square, N.Y. City. M.I. June 30, 1861.

Commanders: Col. David J. Nevin. b. York, Penn., 1828. Coal merchant in N.Y. City. Capt., Co. D, 62nd N.Y., June 30, 1861. M.O. June 29, 1864. d. N.Y. City, Oct. 24, 1880. On July 1, when Nevin commanded brigade, Lt. Col. Theodore Burns Hamilton commanded regiment. b. New York City, 1836. Father Frank H. Hamilton was a prominent military surgeon. Theodore was a law student in Albany, N.Y. at the start of the war. Capt., 33rd N.Y. Lt. Col., 62nd N.Y., Dec. 27, 1862. Wd. May 9, 1864. M.O. Aug. 30, 1865. Member of brokerage firm. d. Queens, N.Y., Nov. 23, 1893.

Number: 237   **Loss:** 1 k., 11 wd.

Monument: North of Wheatfield Road. Map reference **V B-11**

Bas relief represents regiment driving advancing enemy to recapture two Federal cannon.

"July 2, 1863. 7:15 p.m."

63rd NEW YORK INFANTRY
2 Companies, AB
2nd Corps, 1st Div., 2nd Brig.

Other Names: "Third Regiment Irish Volunteers"

Raised: New York City and Albany County.

Organized: Camp Carrigan, Quarantine Grounds, Staten Island, N.Y. M.I. Sept.-Nov., 1861.

Commanders: Lt. Col. Richard Charles Bentley. b. Albany, N.Y., 1830. Commission and shipping merchant in Albany, N.Y. Officer in militia. Adjt., 30th N.Y. Maj., 63rd N.Y., Feb. 14, 1862. Wd. May 3, 1863 and at Gettysburg, July 2, 1863. M.O. Sept. 18, 1864. d. Albany, N.Y., Dec. 1, 1871. When Bentley wd., Capt. Thomas Touhy took command. b. Clare, Ireland, 1833. 2 Lt., Co. A,

63rd N.Y., Aug. 7, 1861. Mort. wd. May 5, 1864 when Maj. of the regiment. d. Brooklyn, N.Y., May 30, 1864.
Number: 112 **Loss:** 5 k., 10 wd., 8 m.
Marker: Sickles Ave. Small flat stone. Map reference **V B-6**

64th NEW YORK INFANTRY
2nd Corps, 1st Div., 4th Brig.

Other Names: "First Cattaraugus Regiment"
Raised: Counties of Cattaraugus, Allegany, and Tompkins.
Organized: Camp Chemung, Elmira, N.Y. M.I. Dec. 10, 1861.
Commanders: Col. Daniel G. Bingham. b. Riga, N.Y., Jan. 30, 1827. Lawyer in Ellicottville, N.Y. and Le Roy, N.Y. Militia officer. Lt. Col., 64th N.Y., Nov. 20, 1861. Wd. June 1, 1862 and July 2, 1863. M.O. Feb. 10, 1864. d. Le Roy, N.Y., July 21, 1864. When Bingham wd., Maj. Leman W. Bradley took command. b. Sharon, Conn., Mar. 6, 1820. Cutlery dealer in Hudson, N.Y. Lt., 14th N.Y. Lt., Co. H, 64th N.Y. Dec. 31, 1861. Wd. June 1, 1862 and May 12, 1864. M.O. Lt. Col., Oct. 5, 1864. d. Hudson, N.Y., Aug. 13, 1912.
Number: 221 **Loss:** 15 k., 64 wd., 19 m.
Monument: Brooke Ave. Map reference **V D-5**
"July 2, 1863."
Marker: Located about 500 feet east of regimental monument. Locates where Capt. Henry V. Fuller was killed on July 2, 1863. Map reference **V D-5**

65th NEW YORK INFANTRY
6th Corps, 3rd Div., 1st Brig.

Other Names: "United States Chasseurs"
Raised: New York City, Seneca County, Oh., and Providence County, R.I.
Organized: Camp Tompkins, Willett's Point, Long Island, N.Y. M.I. July and Aug., 1861.
Commander: Col. Joseph Eldridge Hamblin. b. Yarmouth, Mass., Jan. 13, 1828. Insurance broker. Militia service. Resided in Missouri before the war. "Conspicuous in the Kansas border troubles." Returned to N.Y. City to enlist. Lt., 5th N.Y. Maj., 65th N.Y., Nov. 4, 1861. Wd. Oct. 19, 1864. M.O. Brig. Gen., Jan. 15, 1866. d. N.Y. City, July 3, 1870.
Number: 319 **Loss:** 4 k., 5 wd.
Monument: Slocum Ave. Map reference **III F-5**
"Arrived on the field at 2 p.m. July 2. At daylight of the 3d, moved from base of Little Round Top to Culp's Hill. Held this position till 3 p.m. then moved to left centre."

66th NEW YORK INFANTRY
2nd Corps, 1st Div., 3rd Brig.

Other Names: "Governor's Guard"
Raised: New York City

Organized: Elm Park, N.Y. City. M.I. Nov. 4, 1861.

Commanders: Col. Orlando Harriman Morris. b. 1835. Grad. Columbia College, N.Y., 1854. Lawyer in N.Y. City. Maj., 66th N.Y., Oct. 29, 1861. Wd. July 2, 1863 while carrying the colors. Killed at Cold Harbor, Va., June 3, 1864. When Morris wd., Lt. Col. John Sweeney Hammell took command. b. Trenton, N.J., 1842. Merchant in N.Y. City at start of war. Militia service. Lt., Co. I, 66th N.Y., Sept. 6, 1861. Wd. Dec. 13, 1862 and at Gettysburg. M.O. March 9, 1865. Moved to Montana. d. Camp Baker near Diamond City, Montana, Jan. 31, 1873. Maj. Peter Adolph Nelson took command after Hammell. b. Copenhagen, Denmark. Builder from Westchester, N.Y. Capt., Co. G, 66th N.Y., Sept. 2, 1861. M.O. May 5, 1865. Carpenter in Lynn, Mass. where he died July 23, 1899, age 81 years, 4 months, 17 days.

Number: 176   **Loss:** 5 k., 29 wd., 10 m.

Monument: Sickles Ave. Map reference **V B-6**

"July 2, 1863. 6 p.m."

67th NEW YORK INFANTRY
6th Corps, 3rd Div., 1st Brig.

Other Names: "First Long Island"

Raised: Counties of Kings and Monroe.

Organized: Camp Plymouth, South Brothers Island, East River, New York City. M.I. June 20, 1861.

Commander: Col. Nelson Cross. b. Lancaster, N.H., Feb., 1824. Brother Edward was killed at Gettysburg leading a brigade. Nelson served during the Mexican War. Lawyer in Cincinnati, Oh., member of state legislature, and Judge of Court of Common Pleas for Hamilton County, Oh. Start of the war he was a lawyer in Brooklyn, N.Y. Lt. Col., 67th N.Y., June 24, 1861. M.O. July 4, 1864. d. Mar. 12, 1897 in Dorchester, Mass.

Number: 356   **Loss:** 1 m.

Monument: Slocum Ave. Map reference **III F-5**

"Held this position July 3; then moved double quick to left centre to resist Confederate charge upon our batteries."

68th NEW YORK INFANTRY
11th Corps, 1st Div., 1st Brig.

Other Names: "Cameron Rifles"

Raised: New York City

Organized: Camp Cameron, U.S. Arsenal, Hudson City, N.J. M.I. Aug. 24, 1861.

Commander: Col. Gotthilf Von Bourry d'Ivernois. b. 1822 or 1823. Served in Austrian army beginning 1839. Joined Gen. Louis Blenker's staff Oct. 16, 1861. Col., 68th N.Y., Aug. 6, 1862. Cashiered for drunkenness and neglect of duty, Oct. 25, 1863. Disability removed May 17,1864. No further military service indicated after Oct. 1863.

Number: 264   **Loss:** 8 k., 63 wd., 67 m.

Monument: Wainwright Ave. Map reference II L-3

"This regiment having participated in the first day of the battle, held this position on the 2d and 3d of July, 1863."

69th NEW YORK INFANTRY
2 Companies, AB
2nd Corps, 1st Div., 2nd Brig.

Other Names: "First Regiment Irish Brigade"

Raised: New York City

Organized: Fort Schuyler, Throgs Neck, N.Y. City. M.I. Sept. 7 - Nov. 17, 1861.

Commanders: Capt. Richard Moroney. b. Lockport, N.Y., 1828 or 1829. Mexican War service. Machinist in N.Y. City. Member of 69th N.Y. State Militia. Lt., Co. F, 69th N.Y. Inf., Oct. 12, 1861. Wd. July 2, 1863, M.O. Maj., June 30, 1865. d. Dec. 29, 1865. Lt. James Joseph Smith took command after Moroney wd. b. County Monaghan, Ireland, Sept. 27, 1835. Plumber in N.Y. City. Member 69th N.Y. State Militia. Lt. and Adjt., 69th N.Y. Inf., Nov. 8, 1861. Wd. Aug. 14 and 26, 1864. M.O. Lt. Col., June 30, 1865. d. Oct. 7, 1913 in Cleveland, Oh.

Number: 75   **Loss:** 5 k., 14 wd., 6 m.

Monument: None. Regiment honored on the New York Irish Brigade Monument. Sickles Ave.

Marker: Near brigade monument. Map reference V B-5

70th NEW YORK INFANTRY
3rd Corps, 2nd Div., 2nd Brig.

Other Names: "First Regiment Excelsior"

Raised: New York City. Counties of Van Buren, Mich.; Allegheny, Penn.; Orange, N.Y.; Suffolk, Mass.; Essex, N.J.; and Passaic, N.J.

Organized: Camp Scott, Staten Island, N.Y. as part of Gen. Daniel Sickles' Excelsior Brigade. M.I. June 20, 1861.

Commander: Col. John Egbert Farnum. b. N.J., April 1, 1824. Raised in Pottsville, Penn. Mexican War service. Participant in the Lopez expedition to Cuba in 1850 and Walker's Nicarauguan expedition. Later Captain of the slave ship "Wanderer." Indicted in Savannah, Ga. courts for carrying on the slave trade. Maj., 70th N.Y., June 27, 1861. Wd. May 5, 1862. M.O. July 1, 1864. Col., Veteran Reserve Corps until June 30, 1866. d. New York City, May 16, 1870.

Number: 371   **Loss:** 20 k., 93 wd., 4 m.

Monument: No unit monument. Regiment honored on brigade monument, Sickles Ave.

Marker: Small stone near brigade monument. Map reference IV K-2

71st NEW YORK INFANTRY
3rd Corps, 2nd Div., 2nd Brig.

Other Names: "Second Regiment Excelsior"

Raised: New York City and Philadelphia, Penn. Also counties of Essex, N.J.; Ulster, N.Y.; and Cattaraugus, N.Y.

Organized: Camp Scott, Staten Island, N.Y. as part of Gen. Daniel Sickles' Excelsior Brigade. M.I. June 2 - July 18, 1861.

Commander: Col. Henry Langdon Potter. b. Tyringham, Mass., Mar. 26, 1828. 1850s was a prominent paper manufacturer in Housatonic, Mass. Lawyer at start of war. Lt. Col., 71st N.Y., July 18, 1861. Wd. Aug. 27, 1862 and at Gettysburg, July 2, 1863. M.O. Dec. 31, 1864. Lawyer in Linden, N.J. where he died Mar. 29, 1907.

Number: 243 **Loss:** 10 k., 68 wd., 13 m.

Monument: No unit monument. Regiment honored on the brigade monument, Sickles Ave.

Marker: Small stone near brigade monument. Map reference **IV J-3**

72nd NEW YORK INFANTRY
3rd Corps, 2nd Div., 2nd Brig.

Other Names: "Third Regiment Excelsior"

Raised: New York City. Counties of Chautauqua, N.Y.; Delaware, N.Y.; and Essex, N.J.

Organized: Camp Scott, Staten Island, N.Y. as part of Gen. Sickles' Excelsior Brigade. M.I. June 20, 1861.

Commanders: Col. John S. Austin. b. N.Y., 1817. Start of war was clerk in N.Y. City. Capt., Co. K, 72nd N.Y., June 21, 1861. Wd. July 2, 1863. Discharged for disability June 24, 1864. d. before Sept. 27, 1865. Lt. Col. John Leonard took command after Austin was wounded. b. County Caven, Ireland, Dec. 11, 1835. Occupation hatter. Pvt., 9th U.S. Infantry, 1855-1860. Residence then in Newark, N.Y. Militia officer. Capt., Co. F, 72nd N.Y., June 21, 1861. M.O. June 19, 1864. Service with Veteran Reserve Corps and the 43rd and 1st U.S. Infantry until retirement Dec. 15, 1870. Superintendent of Newark Hospital for the insane. d. Newark, N.J., Feb. 26, 1902.

Number: 366 **Loss:** 7 k., 79 wd., 28 m.

Monument: No unit monument. Regiment honored on Brigade monument, Sickles Ave.

Marker: Small stone near brigade monument. Map reference **IV J-3**

73rd NEW YORK INFANTRY
3rd Corps, 2nd Div., 2nd Brig.

Other Names: "Fourth Regiment Excelsior." "Second Fire Zouaves."

Raised: New York City and Kings County.

Organized: Camp Decker, Staten Island, N.Y. as part of Gen. Sickles' Excelsior Brigade. M.I. July 8 - Oct. 8, 1861. Raised from N.Y. City Fire Companies. On Jan. 20, 1863, 6 companies of the 163rd N.Y. were consolidated with the 73rd N.Y. Inf.

Commander: Maj. Michael William Burns. b. Ireland, 1834. Start of war was city inspector and fireman in N.Y. City. Capt., Co. A, 73rd N.Y., Aug. 14, 1861. Wd. Aug. 27, 1862. M.O. Lt. Col., June 29, 1865. Brought up on charges 3 times for misconduct during military service. Last years served as harbor master in N.Y. City where he died Dec. 7, 1883.

Number: 507 **Loss:** 51 k., 103 wd., 8 m.

Monument: Sickles Ave. Map reference **IV J-2**
"The Fourth Excelsior Regiment was conducted to this position by Major H. E. Tremain, of Third Corps staff about 5:30 p.m., on July 2, 1863."

Marker: Small stone near monument. Map reference **IV K-2**

74th NEW YORK INFANTRY
3rd Corps, 2nd Div., 2nd Brig.

Other Names: "Fifth Regiment Excelsior"

Raised: Allegheny and Warren Counties in Penn., New York City, and Middlesex County, Mass.

Organized: Camp Scott, Staten Island, N.Y. as part of Gen. Sickles' Excelsior Brigade. M.I. June 30 - Oct. 6, 1861.

Commander: Lt. Col. Thomas Holt. b. Manchester, England, Aug. 6, 1831. Carriage maker in Middletown, N.Y. Capt., 70th N.Y. Transferred to command 74th N.Y., May 16, 1863. M.O. Col., 70th N.Y., July 1, 1864. d. Waterbury, Conn., July 9. 1897.

Number: 275 **Loss:** 12 k., 74 wd., 3 m.

Monument: No unit monument. Regiment honored on brigade monument, Sickles Ave.

Marker: Small stone near brigade monument. Map reference **IV 1-3**

76th NEW YORK INFANTRY
1st Corps, 1st Div., 2nd Brig.

Other Names: "Cortland County Regiment"

Raised: Counties of Cortland, Otsego, and Albany.

Organized: Industrial School Barracks, Albany, N.Y. M.I. Oct. 1861.

Commanders: Maj. Andrew Jackson Grover. b. West Dryden, N.Y., Dec. 22, 1830. Orphaned at age 7. Mexican War service. Ordained Methodist Ministry, 1852. Grover had a church in Cortlandville, N.Y. Capt., Co. A, 76th N.Y., Oct. 8, 1861. Wd. Aug. 28, 1862. Killed July 1, 1863. Then Maj. John Elihu Cook took command. b. Hadley, Mass., Aug. 25, 1829. Carpenter in Middleburg, N.Y. Belonged to militia. Capt., Co. I, 76th N.Y., Oct. 14, 1861. Wd. May 5, 1864 and Oct. 7, 1864. M.O. Lt. Col., Oct. 15, 1864. d. April 4, 1899. Buried Bergen County, N.J.

Number: 375 **Loss:** 32 k., 132 wd., 70 m.

Monument: Reynolds Ave. Map reference **I E-6**

"Fire opened here, July 1, 1863, at 10 a.m. Second stand at R.R. cut. Third, at Culp's Hill July 2d and 3d."
Marker: Culp's Hill. Map reference III E-2

77th NEW YORK INFANTRY
6th Corps, 2nd Div., 3rd Brig.

Other Names: "Bemis Heights Regiment"
Raised: Counties of Saratoga, Essex and Fulton.
Organized: Camp Schuyler, Saratoga Springs, N.Y. M.I. Nov. 23, 1861.
Commander: Lt. Col. Winsor Brown French. b. Cavendish, Vt., July 28, 1832. Grad. Tufts College, Mass., 1859. Lawyer in Saratoga Springs, N.Y. Lt. and Adjt., 77th N.Y., Sept. 24, 1861. Wd. July 13, 1864. M.O. Dec. 13, 1864. District Attorney for Saratoga County. d. Mar. 24, 1910 in Saratoga Springs, N.Y.
Number: 424   **Loss:** None
Monument: Powers Hill. Map reference III F-18
   "July 3, 1863."

78th NEW YORK INFANTRY
12th Corps, 2nd Div., 3rd Brig.

Other Names: "Cameron Highlanders"
Raised: New York City. Counties of Erie, Steuben, Oneida, Wyoming, Monroe and Niagara.
Organized: Tompkinsville, Staten Island, N.Y. M.I. Oct. 1, 1861 - April, 1862.
Commander: Lt. Col. Herbert Von Hammerstein. b. Hanover, Germany, Dec. 20, 1835. Austrian officer. Start of war gives residence Washington, D.C. Capt., Co. A, 8th N.Y., April 23, 1861. Aug. 15, 1861 joined staff of Gen. McClellan. Lt. Col., 78th N.Y., June 16, 1863. M.O. Jan. 7, 1865. Sergt., 2nd U.S. Cav., July 17, 1865 to August 14, l867. Legs were frozen in cold weather to a point where amputation required. Returned to Europe.
Number: 198   **Loss:** 6 k., 21 wd., 3 m.
Monument: Slocum Ave. 78th N.Y. and 102 N.Y. combined.
   Map reference III F-4
   "Ground occupied during the battle by 102 N.Y. Regt. and 78 N.Y. Skirmishers on ground in front."

80th NEW YORK INFANTRY
1st Corps, 3rd Div., 1st Brig.

Other Names: "Ulster Guard." "Twentieth New York State Militia."
Raised: Ulster County
Organized: Camp Arthur near Kingston, N.Y. M.I. Sept. - Oct., 1861.
Commander: Col. Theodore Burr Gates. b. Oneonta, N.Y., Dec. 16, 1825. Lawyer from Kingston, N.Y. Belonged to Twentieth N.Y. State Militia. Lt. Col., 80th N.Y., Sept. 10, 1861. M.O. Nov. 22, 1864. Lawyer in Brooklyn, N.Y. where he died July 5, 1911.

Number: 375 **Loss:** 35 k., 111 wd., 24 m.
Monument: Reynolds Ave. Map reference **I J-4**
"Held substantially this position from about 12 m. July 1, 1863 to 4 p.m. July 2 on Cemetery Hill in support of 3d Corps. July 3rd in front line of battle resisting Pickett's attack."
Marker: Hancock Ave. Map reference **II E-15**
Position of July 3, 1863.

82nd NEW YORK INFANTRY
2nd Corps, 2nd Div., 1st Brig.

Other Names: "Second New York State Militia"
Raised: New York City
Organized: Camp Anderson on the Battery, N.Y. City. M.I. June 17, 1861.
Commanders: Lt. Col. James Huston. b. Ireland, Jan. 7, 1818. Clerk in N.Y. City. Capt., Co. E, 82nd N.Y., May 21, 1861. Killed July 2, 1863 at Gettysburg. Capt. John Darrow. b. N.Y. City, 1821 or 1822. Cooper in N.Y. City. Capt., Co. K, 82nd N.Y., May 21, 1861. M.O. Lt. Col., Sept. 14, 1863.
Number: 394 **Loss:** 45 k., 132 wd., 15 m.
Monument: Hancock Ave. Map reference **II E-14**
"On the evening of July 2 moved to the Emmitsburg road to protect flank of Third Corps. Fought there until out-flanked. Returning to this line the regiment reformed under a galling fire; then advanced, driving the enemy before them; regained their former position, capturing the colors of the 48th Georgia. At the time of the enemy's assault on the afternoon of the 3d, the regiment moved to the right toward the Copse of Trees and assisted in repulsing the enemy, capturing the flags of the First and Seventh Virginia Regiments."

83rd NEW YORK INFANTRY
1st Corps, 2nd Div., 2nd Brig.

Other Names: "Ninth Militia." "City Guards."
Raised: New York City
Organized: Ninth Militia Armory, N.Y. City. Drilled in Washington Square. M.I. June 8, 1861.
Commander: Lt. Col. Joseph Anton Moesch. b. Eiken, Canton Aagau, Switzerland, August 13, 1829. Came to U.S. in 1854. Baker and clerk in N.Y. City. Sergt., Co. B, 83rd N.Y., April 29, 1861. Killed as Col. of the regiment on May 6, 1864 at the Wilderness, Va. Buried Fredericksburg National Cemetery.
Number: 215 **Loss:** 6 k., 18 wd., 58 m.
Monument: Doubleday Ave. Map reference **I D-7**
"Engaged on this ground July 1, 1863, 1 p.m. to 3 p.m. assisting in capture of Iverson's N.C. Brigade, C.S.A. July 2 and 3, 1863 at Ziegler's Grove, also supported Batteries with llth and 2d Corps."

84th NEW YORK INFANTRY
1st Corps, 1st Div., 2nd Brig.

Other Names: "Fourteenth Militia." "Brooklyn Chasseurs."
Raised: Kings County
Organized: Fort Greene, Brooklyn, N.Y. M.I. May and Aug., 1861.
Commander: Col. Edward Brush Fowler. b. N.Y. City, May 29, 1828. Book-keeper in Brooklyn. Lt. Col., 84th N.Y., May 23, 1861. Wd. Aug. 29, 1862. M.O. June 6, 1864. d. Brooklyn, Jan. 16, 1896.
Number: 356 **Loss:** 13 k., 105 wd., 99 m.
Monument: Reynolds Ave. Map reference I G-5
"July 1. First engaged the enemy between the McPherson House and Reynolds Grove [see marker]; subsequently moved to this place and engaged Davis' Brigade; remained at the railroad cut at Seminary Ridge until the final retreat; had a running fight through Gettysburg to Culp's Hill. On the evening of the 2d and again on the morning of the 3d went to support Greene's Brigade and was heavily engaged [see marker]."
Markers: Stone Ave. and Slocum Ave. (attached to boulder).
Map references I H-3, III F-6

86th NEW YORK INFANTRY
3rd Corps, 1st Div., 2nd Brig.

Other Names: "Steuben Rangers"
Raised: Counties of Steuben and Chemung.
Organized: Elmira, N.Y. M.I. Nov. 23, 1861.
Commanders: Lt. Col. Benjamin L. Higgins. b. Brewster, Mass., Oct. 14, 1826. Chief engineer of Syracuse, N.Y. fire dept. Capt., Co. A, 86th N.Y., Nov. 12, 1861. M.I. Col., July 1, 1863. Wd., July 2, 1863 and Nov. 27, 1863. M.O. June 25, 1864. Wholesale liquor business. d. Syracuse, N.Y., Nov. 19, 1891. Maj. Jacob H. Lansing took command when Higgins wd. b. Albany, N.Y., 1824. Jeweler in Corning, N.Y. Capt., Co. C, 86th N.Y., Nov. 12, 1861. Lt. Col., July 1, 1863. Wd. May 24, 1864. M.O. Nov. 14, 1864. Served in Corning city government. d. Corning, Nov. 8, 1885.
Number: 286 **Loss:** 11 k., 51 wd., 4 m.
Monument: Sickles Ave. Map reference V D-7
"This regiment held this position the afternoon of July 2, 1863."

88th NEW YORK INFANTRY
2 Companies, AB
2nd Corps, 1st Div., 2nd Brig.

Other Names: "Fifth Regiment Irish Brigade"
Raised: New York City
Organized: Fort Schuyler, Throgs Neck, N.Y. City. M.I. Sept. 61 - Jan. 62.
Commander: Capt. Denis Francis Burke. b. Cork, Ireland, April 19, 1841. Dry

goods business in N.Y. City. 2 Lt., Co. C, Dec. 11, 1861. Wd. Dec. 13, 1862 and May 3, 1863. M.O. Lt. Col., June 30, 1865. Publisher in N.Y. City. d. N.Y. City, Oct. 19, 1893.

Number: 126   **Loss:** 7 k., 17 wd., 4 m.

Monument: No unit monument. Regiment honored on the New York Irish Brigade Monument, Sickles Ave.

93rd NEW YORK INFANTRY

Only detachments on the field at Gettysburg.

94th NEW YORK INFANTRY
1st Corps, 2nd Div., 1st Brig.

Other Names: "Belle Jefferson Rifles"

Raised: Jefferson County

Organized: Madison Barracks, Sacket's Harbor, N.Y. M.I. Dec. 9, 1861. Formed from consolidation of 94th N.Y. and 105th N.Y. regiments.

Commanders: Col. Adrian Rowe Root. b. Buffalo, N.Y., May 6, 1833. Provision dealer in Buffalo and member of Buffalo City Guards. Lt. Col., 21st N.Y. inf. Col., 94th N.Y., May 2, 1862. Wd. Second Bull Run and July 1, 1863. M.O. July 18, 1865. d. June 4, 1899 in Buffalo. When Root wd., Maj. Samuel A. Moffett took command. b. Rodman, N.Y., July 4, 1836. Clerk in Rodman. Pvt., Co. A, 94th N.Y., Sept. 27, 1861. M.O. Lt. Col., July 18, 1865. d. Ridgeland, Mississippi, Mar. 24, 1917.

Number: 445   **Loss:** 12 k., 58 wd., 175 m.

Monument: Doubleday Ave. Map reference I E-7
"July 1, 1863."

95th NEW YORK INFANTRY
1st Corps, 1st Div., 2nd Brig.

Other Names: "Warren Rifles"

Raised: New York City. Counties of Westchester, Rockland and Schoharie.

Organized: Camp Lafayette, New Dorp, Staten Island, N.Y. M.I. Nov. 1861 - Mar. 1862.

Commanders: Col. George H. Biddle. b. N.Y. City, Oct. 1, 1802. Mexican War service. Militia service. Clerk, N.Y. City. Col., 95th N.Y., Dec. 20, 1861. Wd. Gettysburg. M.O. Oct. 9, 1863. d. N.Y. City, June 11, 1884. Maj. Edward Pye took command at Gettysburg after Biddle wd. b. Rockland County, N.Y., Sept. 5, 1823. Grad. Rutgers College, N.J. Lawyer, Haverstraw, N.Y. Capt., Co. F, 95th N.Y., Oct. 15, 1861. Wd. June 2, 1864 at Cold Harbor, Va. d. of wounds June 12, 1864.

Number: 261   **Loss:** 7 k., 62 wd., 46 m.

Monument: Reynolds Ave. Map reference I G-5
"July 1, 1863. This regiment was formed south of the McPherson House and

engaged the enemy at 10 a.m. At 10:30 a.m. changed front, advanced to this position with the 84th New York and 6th Wisconsin, repulsed and captured a large part of Davis' Mississippi Brigade in the railroad cut. At noon, July 1st, held position on Oak Hill indicated by marker; being outflanked moved to the right of Seminary supporting Battery B, 4th U.S. Retired from that position to Culp's Hill, where it remained during July 2 and 3."

Markers: Stone Ave. (10 a.m.), Wadsworth Ave. (12 n.), Seminary Ave. at U.S. Rt. 30 (4 p.m.), Culp's Hill (July 2 and 3).

Map references I G-3, I F-7, I H-6, III F-3

97th NEW YORK INFANTRY
1st Corps, 2nd Div., 2nd Brig.

Other Names: "Conkling Rifles"
Raised: Counties of Oneida, Herkimer and Lewis.
Organized: Camp Rathbone, Boonville, N.Y. M.I. Feb. 1862.
Commanders: Col. Charles Wheelock. b. Claremont, N.H., Dec. 14, 1812. Farmer and produce dealer in Boonville, N.Y. Militia officer. Col., 97th N.Y., Feb. 7, 1862. Wd. and captured on July 1, 1863. d. of disease Jan. 21, 1865 in Washington, D.C. When Wheelock wd., Maj. Charles B. Northup took command. b. Deerfield, N.Y., May 1, 1828. Left job in Oneida Central Bank in Rome, N.Y. to join Co. K, 97th N.Y. as pvt., Oct. 6, 1861. Wd. May 6, 1864. M.O. Dec. 9, 1864. Bank employee in Chicago, Ill. d. Chicago, Jan. 28, 1918.
Number: 255 **Loss:** 12 k., 36 wd., 78 m.
Monument: Doubleday Ave. Map reference I D-7
"Held the enemy in check here from 12:30 to 3 p.m. July 1, 1863. During this time charged across the field to the west, assisting in capturing Iverson's Brigade, and securing flag of 20th N.C."

102nd NEW YORK INFANTRY
12th Corps, 2nd Div., 3rd Brig.

Raised: New York City. Counties of Ulster and Schoharie.
Organized: New Lots, Kings County, N.Y. M.I. Sept. 1861 - April 1862.
Commanders: Col. James Crandall Lane b. N.Y. City, July 28, 1823. Civil Engineer. When Civil War began engaged in mineralogical surveys in Santo Domingo, Puerto Rico and Cuba. Maj., 102nd N.Y., Jan. 1, 1862. Wd. July 2, 1863. M.O. July 12, 1864. Directed mineralogical surveys in California and archaeological surveys in Palestine and the river Jordan. d. Brooklyn, N.Y., Dec. 13, 1888. Capt. Lewis R. Stegman took command when Lane wounded. b. N.Y. City, Jan. 18, 1840. Law student in N.Y. City. Capt., Co. E, 102nd N.Y., Mar. 5, 1862. Wd. Aug. 9, 1862 and June 16, 1864. M.O. Oct. 24, 1864. Service in U.S. Veteran Volunteers until Feb. 20, 1866. d. Brooklyn, N.Y., Oct. 7, 1923.
Number: 248 **Loss:** 4 k., 17 wd., 8 m.

Monument: 78th N.Y. and 102nd N.Y. combined. Slocum Ave.
Map reference III F-4
See 78th N.Y. for inscription.

104th NEW YORK INFANTRY
1st Corps, 2nd Div., 1st Brig.

Other Names: "Wadsworth Guards"
Raised: Counties of Livingston and Rensselaer.
Organized: Albany, N.Y. M.I. Sept. 1861 - Mar. 1862.
Commander: Col. Gilbert G. Prey. b. New Brunswick (Canada), 1822. Carpenter and joiner in Eagle, N.Y. Capt., Co. F, Dec. 5, 1861. M.O. Mar. 3, 1865. d. Eagle, N.Y., Feb. 6, 1903.
Number: 309   **Loss:** 11 k., 91 wd., 92 m.
Monument: Robinson Ave. Map reference I C-8
"July 1, 1863."

105th NEW YORK INFANTRY

See 94th N.Y. Infantry

107th NEW YORK INFANTRY
12th Corps, 1st Div., 3rd Brig.

Other Names: "Campbell Guards"
Raised: Counties of Chemung and Steuben.
Organized: Elmira, N.Y. M.I. Aug. 13, 1862.
Commander: Col. Nirom Marium Crane. b. Penn Yan, N.Y. Dec. 13, 1828. Banker in N.Y. City living in Hornellsville, N.Y. Militia officer. Lt. Col., 23rd N.Y. Col., 107th N.Y., June 24, 1863. M.O. June 5, 1865. d. Wayne, N.Y., Sept. 19, 1901.
Number: 319   **Loss:** 2 wd.
Monument: Slocum Ave. Map reference III H-9
"Occupied this position on the morning of July 2. Withdrawn at 7 p.m., and sent to near Little Round Top. Returning during the night found these works in possession of the enemy. During the morning of July 3, was in position near Baltimore Pike. Reoccupied these works about 2 p.m."

108th NEW YORK INFANTRY
2nd Corps, 3rd Div., 2nd Brig.

Other Names: "Rochester Regiment"
Raised: Monroe County
Organized: Camp Fitz John Porter, near Rochester, N.Y. M.I. Aug. 16-18, 1862.
Commander: Col. Francis Edwin Pierce. b. Fowlerville, N.Y., July 6, 1833. Grad. Rochester University, N.Y., 1859.   Established Rochester Military Academy. Capt., Co. F, 108th N.Y., Aug. 18, 1862. Wd. Feb. 14, 1864 and May

6, 1864. M.O. May 28, 1865. Served in Veteran Volunteers and 1st U.S. Infantry to Jan. 8, 1880. d. at the Presidio in San Francisco, California, Nov. 4, 1896. Buried San Francisco National Cemetery.

Number: 305   **Loss:** 16 k., 86 wd.

Monument: Hancock Ave. Map reference **II F-8**

"Occupied this position July 2 and 3,1863, supporting Battery I, 1st U.S. Art. During the artillery duel on the afternoon of July 3, it sustained a terrific fire without being able to return a shot."

111th NEW YORK INFANTRY
8 Companies (Companies B and C absent)
2nd Corps, 3 Div., 3rd Brig.

Raised: Counties of Cayuga and Wayne.

Organized: Auburn, N.Y. M.I. Aug. 20, 1862.

Commanders: Col. Clinton Dugald MacDougall. b. Kintyre, Scotland, June 14, 1839. Banker in Auburn, N.Y. Capt., 75th N.Y. Lt. Col., 111th N.Y., Aug. 20, 1862. Wd. July 3, 1863 and April 2, 1865. M.O. June 4, 1865. Elected to U.S. Congress, 1873-1877. d. Paris, France, May 24, 1914. Buried Arlington National Cemetery. After MacDougall, Lt. Col. Isaac M. Lusk took command. b. Newark, N.Y., 1826. Farmer in Newark. Lt., 17th N.Y. Inf. Capt., Co. E, 111th N.Y., Aug. 15, 1862. Wd. July 3, 1863. M.O. April 2, 1864. Farmer in Petoskey, Mich. d. May 4, 1908. Capt. Aaron Platt Seeley took command from Lusk. b. Easton, Conn., Nov. 8, 1832. Carriage maker, Palmyra, N.Y. Capt., Co. A, 111th N.Y., Aug. 5, 1862. Wd. May 5, 1864. M.O. Aug. 19, 1864. d. Palmyra, Dec. 30, 1920.

Number: 390 (8 companies)   **Loss:** 58 k., 177 wd., 14 m.

Monument: Hancock Ave. Map reference **II F-9**

"Arrived early morning July 2, 1863, position near Ziegler's Grove. Went to relief of 3d Corps in afternoon; took this position that evening and held it to close of battle."

119th NEW YORK INFANTRY
11th Corps, 3rd Div., 2nd Brig.

Raised: New York City and Queens County.

Organized: Camp Peissner, Turtle Bay Park, near 45th Street, N.Y. City. M.I. Sept. 4 and 5, 1862.

Commanders: Col. John Thomas Lockman. b. N.Y. City, Sept. 26, 1834. Law student, N.Y. City. Pvt., 9th N.Y. State Militia. Lt. Col., 119th N.Y., Oct. 16, 1862. Wd. July 1, 1863. M.O. June 7, 1865. Grad Columbia University, N.Y., 1867. d. New York City, Sept. 27, 1912. Lt. Col. Edward F. Lloyd took command from Lockman on July 1, 1863. b. N.Y., Aug. 19, 1836. Connected with map publishing firm in N.Y. City. Lt., Co. F, 119th N.Y., Aug. 8, 1862. Killed in Battle of Resaca, Ga., May 15, 1864. Buried Arlington National Cemetery.

Number: 300   **Loss:** 11 k., 70 wd., 59 m.

Monument: Howard Ave. Map reference **I D-13**
   "July 1, 1863."

120th NEW YORK INFANTRY
3rd Corps, 2nd Div., 2nd Brig.

Other Names: "Washington Guards"
Raised: Counties of Ulster and Greene.
Organized: Camp Samson, Kingston, N.Y. M.I. Aug. 22, 1862. See also 4th
   U.S. Artillery, Battery K.
Commander: Lt. Col. Cornelius Depuy Westbrook. b. Fishkill, N.Y., Jan. 13,
   1823. Grad. Rutgers College, N.J., 1838. Civil Engineer in Kingston, N.Y.
   Capt., 20th N.Y. State Militia. Lt. Col., 120th N.Y., Aug. 22, 1862. Wd. July 2,
   1863. M.O. Feb. 27, 1864. d. Kingston, Sept. 24, 1905. When Westbrook wd.,
   Maj. John Rudolph Tappen took command. b. Kingston, N.Y., May 26, 1831.
   Merchant in Kingston. Capt., 20th N.Y. State Militia. Maj., 120th N.Y., Sept.
   8, 1862. Wd. Aug. 30, 1862. M.O. Lt. Col., Dec. 3, 1864. d. Kingston, Jan. 20,
   1875.
Number: 427   **Loss:** 32 k., 154 wd., 17 m.
Monument: Sickles Ave. Map reference **IV H-4**
   The One Hundred and Twentieth New York Infantry held this part of the
   line on the second day of July, 1863."

121st NEW YORK INFANTRY
6th Corps, 1st Div., 2nd Brig.

Raised: Counties of Otsego and Herkimer.
Organized: Camp Schuyler, German Flats, near Herkimer, N.Y. M.I. Aug. 13,
   1862.
Commander: Col. Emory Upton. b. Batavia, N.Y., Aug. 27, 1839. Attnd. Oberlin
   College, Oh. Grad. West Point, 1861. Artillery service. Col., 121st N.Y., Oct.
   23, 1862. Wd. First Bull Run and Sept. 19, 1864. M.O. volunteer service as
   Brig. Gen., April 30, 1866. Regular service until he shot himself dead at the
   Presidio in San Francisco, California, Mar. 15, 1881. On July 4, 1863 Upton
   wrote, "Lee's attack yesterday was imposing and sublime. For about ten
   minutes I watched the contest, when it seemed that the weight of a hair
   would have turned the scales."
Number: 470 **Loss:** 2 wd.
Monument: Sykes Ave. Map reference **V E-12**
   "Held this position from evening of July 2d, 1863, until close of battle."

122nd NEW YORK INFANTRY
6th Corps, 3rd Div., 1st Brig.

Other Names: "Third Onondaga"
Raised: Onondaga County

81

Organized: Camp Andrews, south side, Syracuse, N.Y. M.I. Aug. 28, 1862.
Commander: Col. Silas Titus. b. Cato, N.Y., May 30, 1811. Militia officer. Lumber dealer in Syracuse. Lt. and Adjt., 12th N.Y. Inf. Col., 122nd N.Y., Aug. 28, 1862. M.O. Jan. 23, 1865. d. in a house fire in Brooklyn, N.Y., Oct. 4, 1899.
Number: 456   **Loss:** 10 k., 32 wd., 2 m.
Monument: Slocum Ave. Map reference **III F-5**
"Assisted in repulsing the attack on the morning of July 3, 1863."

123rd NEW YORK INFANTRY
12th Corps, 1st Div., 1st Brig.

Other Names: "Washington County Regiment"
Raised: Washington County
Organized: Camp Washington, Salem, N.Y., Sept. 4, 1862.
Commanders: Lt. Col. James Clarence Rogers. b. Sandy Hill, N.Y., Oct. 29, 1838. Grad. Union College, N.Y., 1860. Lawyer, Sandy Hill. Capt., 43rd N.Y. Maj., 123rd N.Y., Sept. 30, 1862. M.O. Col., June 8, 1865. d. Sandy Hill, Feb. 9, 1907. Capt. Adolphus Hitchcock Tanner. b. Granville, N.Y., May 23, 1833. Lawyer in Whitehall, N.Y. Capt., Co. C, 123rd N.Y., Aug. 14, 1862. Wd. May 25, 1864. M.O. Lt. Col., June 8, 1865. Elected to U.S. Congress, 1869-1871. d. Whitehall, N.Y., Jan. 14, 1882.
Number: 495   **Loss:** 3 k., 10 wd., 1 m.
Monument: Slocum Ave. Map reference III G-7
"July 1; marched from Littlestown; formed line of battle on Wolf Hill; bivouacked near Baltimore Pike; July 2, advanced to this line and built a heavy breastwork of logs. At about 6 p.m. moved to support the left near Little Round Top; returning in the night, found breastworks in possession of enemy, as no troops were left to occupy them. July 3, at about 11 a.m. made a charge and recovered these works; about 4 p.m. moved to support line then repelling Pickett's charge; a little later had a sharp skirmish in front of this line; at night repelled an attack with heavy loss to the enemy. July 4; made reconnaissance around Wolf Hill and through Gettysburg over the Hanover road."
Marker: East of monument. Map reference **III G-7** Locates skirmish line on July 3, 1863.

124th NEW YORK INFANTRY
3rd Corps, 1st Div., 2nd Brig.

Other Names: "Orange Blossoms"
Raised: Orange County
Organized: Camp Wickham, Goshen, N.Y. M.I. Sept. 5, 1862.
Commanders: Col. Augustus Van Horne Ellis. b. N.Y. City, May 1, 1827. Attnd. Columbia University, N.Y. Sea captain. Start of war lived in New Windsor, N.Y. Capt., 71st N.Y. State Militia. Col., 124th N.Y., Sept. 5, 1862. Killed July 2, 1863. An acquaintance called him, "a rather cold harsh ambitious man, and

sometimes chilled us with his terrible bursts of profanity; but he was every inch a soldier." Lt. Col. Francis Markoe Cummins. b. Orange County, N.Y., June 29, 1822. Mexican War service. Miller in Muscatine, Iowa. Served in 1st Iowa Inf. and 6th Iowa Inf. (dismissed for drunkenness while leading regiment at Shiloh). Lt. Col., 124th N.Y., Aug. 16, 1862. Wd. July 2, 1863 and May 6, 1864. M.O. Col., Sept. 19, 1864. d. Goshen, N.Y., Mar. 26, 1884.

Number: 279   **Loss:** 28 k., 57 wd., 5 m.

Monument: Sickles Ave. Map reference **V E-7**

"July 2, 1863." Monument displays figure of Ellis.

Marker: Pleasonton Ave. Map reference **II F-16**

125th NEW YORK INFANTRY
2nd Corps, 3rd Div., 3rd Brig.

Raised: Rensselaer County

Organized: Camp Halleck, Troy, N.Y. M.I. Aug. 27, 1862.

Commander: Lt. Col. Levin Crandell. b. Crandell's Corners, Easton, N.Y., Dec. 22, 1826. Militia service. Bookkeeper in a Troy, N.Y. bank. Adjt., 24th N.Y. Inf., Lt. Col., 125th N.Y., Aug. 27, 1862. Wd. June 16, 1864. M.O. Col., Nov. 29, 1864. d. Jamaica, N.Y., June 16, 1907. Buried Cypress Hills National Cemetery.

Number: 500   **Loss:** 26 k., 104 wd., 9 m.

Monument: Hancock Ave. Map reference **II F-10**

"July 2, 1863. Regiment in line of the stone wall until 7 p.m., when the brigade went to the support of the Third Corps. Charged and drove back Barksdale's Mississippi Brigade. Returned at 8:30 p.m.

July 3,1863. Regiment in front on line of the stone wall, west side of Hancock Avenue at time of Longstreet's assault."

126th NEW YORK INFANTRY
2nd Corps, 3rd Div., 3rd Brig.

Raised: Counties of Seneca, Yates, and Ontario.

Organized: Camp Swift, Geneva, N.Y. M.I. Aug. 22, 1862.

Commanders: Col. Eliakim Sherrill. b. Greenville, N.Y., Feb. 16, 1813. Militia officer. Elected to U.S. Congress, 1847-1849. Member state senate. Farmer in Geneva. Col., 126th N.Y., Aug. 20, 1862. Wd. Sept. 15, 1862. Mort. wd. July 3, 1863. d. July 4, 1863. Lt. Col. James M. Bull commanded regiment after Sherrill. b. Canandaigua, N.Y., 1825. Lawyer in Canandaigua, N.Y. Lt. Col., 126th N.Y., Aug. 15, 1862. M.O. Col., April 18, 1864. d. Canandaigua, July 25, 1867.

Number: 511   **Loss:** 40 k., 181 wd., 10 m.

Monument: Hancock Ave. Monument shows relief portrait of Sherrill. Map reference **II F-8**

"The regiment was in position two hundred yards at the left, July 2 until 7 p.m., when the brigade was conducted thirteen hundred yards further to the left and the regiment with the 111th N.Y. and 125th N.Y., charged the enemy in the swale, near the source of Plum Run, driving them therefrom and advancing one hundred and seventy five yards beyond, towards the Emmitsburg Road, to a position indicated by a monument on Sickles Avenue. At dark the regiment returned to near its former position. In the afternoon of July 3 it took this position and assisted in repulsing the charge of the enemy, capturing three stands of colors and many prisoners." Monument indicated above on Sickles Avenue is to Willard's brigade.

134th NEW YORK INFANTRY
11th Corps, 2nd Div., 1st Brig.

Raised: Counties of Schoharie and Schenectady.
Organized: Camp Vedder, Schohaire, N.Y. M.I. Sept. 22, 1862.
Commander: Lt. Col. Allan Hyre Jackson. b. Gibson, N.Y., July 24, 1836. Grad. Union College, N.Y. and Harvard University Law School. Militia service. Lawyer in Schenectady, N.Y. Capt., 91st N.Y. Maj., 134th N.Y., Feb. 27, 1863. Wd. July 20, 1864. M.O. June 10, 1865. Joined 7th U.S. Infantry. Retd. Oct. 29, 1898. d. Schenectady, Aug. 22, 1911.
Number: 488   **Loss:** 42 k., 151 wd., 59 m.
Monument: East Cemetery Hill. Map reference **II K-3**
"July 1, 1863, this regiment was engaged about one quarter mile east of Gettysburg near York road. July 2nd and 3rd, held this position."
Marker: Coster Ave. Map reference **I H-14**
Position of July 1 where it suffered greatest casualties of any battle in which it was ever engaged.

136th NEW YORK INFANTRY
11th Corps, 2nd Div., 2nd Brig.

Raised: Counties of Livingston, Allegany and Wyoming.
Organized: Camp Williams, Portageville, N.Y. M.I. Sept. 26, 1862.
Commander: Col. James Wood, Jr. b. Alstead, N.H., April 4, 1820. Grad. Union College, N.Y., 1842. Lawyer in Geneseo, N.Y. Col., 136th N.Y., Sept. 17, 1862. M.O. June 13, 1865. State senator, 1870-1874. d. Dansville, N.Y., Feb. 24, 1892.
Number: 488   **Loss:** 17 k., 89 wd., 3 m.
Monument: Taneytown Road. Map reference **II H-5**
"July 1, 2, 3, 1863."

137th NEW YORK INFANTRY
12th Corps, 2nd Div., 3rd Brig.

Raised: Counties of Broome, Tioga and Tompkins.
Organized: Camp Susquehanna, Binghamton, N.Y. M.I. Sept. 26, 1862.
Commander: Col. David Ireland. b. Scotland, 1832. Tailor in N.Y. City. Lt. and Adjt., 79th N.Y. Capt., 15th U.S. Infantry. Early in 1862 he was recruiting for the U.S. Regulars in Binghamton, N.Y. when he met and, on Aug. 26,

1863, married Sara Phelps, niece of a prominent local judge. Col., 137th N.Y., Sept. 25, 1862. d. of dysentery, Sept. 10, 1864. Buried in Binghamton.
Number: 456 **Loss:** 40 k., 87 wd., 10 m.
Monument: Slocum Ave. Map reference **III F-5**
"The 137th Regiment of New York Infantry held this position July 2d 1863, and until the retreat of the Rebel Army."

140th NEW YORK INFANTRY
5th Corps, 2nd Div., 3rd Brig.

Raised: Monroe County
Organized: Camp Fitz John Porter, Rochester, N.Y. M.I. Sept. 13, 1863.
Commanders: Col. Patrick Henry O'Rorke. b. County Caven, Ireland, Mar. 28, 1836. He learned father's trade of marble cutter. Grad. West Point, 1861. Engineer service. Col., 140th N.Y., Sept. 19, 1862. Killed at Gettysburg, July 2, 1863. Lt. Col. Louis Ernst assumed command after the death of O'Rorke. b. Baden, Germany, July 19, 1825. Hardware merchant in Rochester, N.Y. Lt. Col. 140th N.Y., Sept. 13, 1862. M.O. Aug. 15, 1863. d. Rochester, N.Y., April 3, 1892.
Number: 526 **Loss:** 26 k., 89 wd., 18 m.
Monument: Sykes Ave. Map reference **V F-11**
"July 2 and 3, 1863."

145th NEW YORK INFANTRY
12th Corps, 1st Div., 1st Brig.

Other Names: "Stanton Legion"
Raised: New York City
Organized: Camp New Dorp, New Dorp, Staten Island, N.Y. M.I. Sept. 11, 1862.
Commander: Col. Edward Livingston Price. b. New York City, Dec. 25, 1844. Capt., 74th N.Y. Inf. Col., 145th N.Y., Feb. 4, 1863. Wd. May 2, 1863. M.O. Dec. 20, 1863. d. Newark, New Jersey, Feb. 4, 1922.
Number: 245 **Loss:** 1 k., 9 wd.
Monument: Slocum Ave. Map reference **III G-8**
"July 2, 3, 1863."

146th NEW YORK INFANTRY
5th Corps, 2nd Div., 3rd Brig.

Other Names: "Halleck Infantry." "Fifth Oneida."
Raised: Oneida County
Organized: Camp Huntington, Rome, N.Y. M.I. Oct. 10, 1862.
Commanders: Col. Kenner Garrard. b. "Fairfield," Bourbon County, Kentucky, Sept. 30, 1827. Attnd. Harvard University. Grad. West Point, 1851. Regular military service. Col., 146th N.Y., Sept. 23, 1862. M.O. Brig. Gen., Aug. 24,

1865. Resigned from Regular service Nov. 9, 1866. d. Cincinnati, Oh., May 15, 1879. When Garrard took command of brigade on July 2, 1863, Lt. Col. David Tuttle Jenkins took command of regiment. b. Oneida County, May 4, 1836. Surveyor and lawyer in Vernon, N.Y. Lt. and Adjt., 146th N.Y., Aug. 26, 1862. Mort. wd., Col., May 5, 1864, during the Battle of the Wilderness, Va.

Number: 534 **Loss:** 4 k., 24 wd.

Monument: Sykes Ave. Map reference **V E-11**

"July 2, 3, 1863." "From this position Maj. Gen. Meade observed the battle for a time on July 3."

147th NEW YORK INFANTRY
1st Corps, 1st Div., 2nd Brig.

Other Names: "Oswego Regiment"

Raised: Oswego County

Organized: Oswego, N.Y. M.I. Sept. 22-23, 1862.

Commanders: Lt. Col. Francis Charles Miller. b. Herkimer County, 1830. Carpenter and joiner in Oswego, N.Y. Capt., 24th N.Y. Inf. Maj., 147th N.Y., Oct. 4, 1862. Wd. July 1, 1863 and May 5, 1864. M.O. Col., June 7, 1865. d. Oneida, N.Y., Aug. 17, 1878. When Miller wd., Maj. George Harney took command. b. Tipperary, Ireland. Shoemaker. Joined Regular Army at Boston in 1852. Capt., Co. B, 147th N.Y., Sept. 22, 1862. Wd. May 10 and Aug. 19, 1864. M.O. Lt. Col., June 7, 1865. Miner in Colorado. d. a hermit living on Leavenworth Mtn. near Georgetown, Colo., Nov. 23, 1881, age 45 years, 3 months, 14 days.

Number: 430 **Loss:** 60 k., 144 wd., 92 m.

Monument: Reynolds Ave. Map reference **I F-5**

"Position 10 a.m. July 1, 1863."

Marker: Slocum Ave. Map reference **III F-5**

"Position 10 a.m. July 1, 1863."

149th NEW YORK INFANTRY
12th Corps, 2nd Div., 3rd Brig.

Other Names: "Fourth Onondaga"

Raised: Onondaga County

Organized: Camp White, Syracuse, N.Y. M.I. Sept. 18, 1862.

Commanders: Col. Henry Alanson Barnum. b, Jamesville, N.Y., Sept. 24, 1833. Grad. Syracuse Institute. Syracuse lawyer and militia member. Capt., 12th N.Y. Inf. Col., 149th N.Y., Sept. 17, 1862. Wd. Malvern Hill and Nov. 23, 1863. M.O. Brig. Gen., Jan. 9, 1866. Awarded Medal of Honor for action at Lookout Mtn., Tenn., Nov. 23, 1863. d. N.Y. City, Jan. 29, 1892. Lt. Col. Charles Bertrand Randall took command from an ill Barnum on night of 2nd. b. Arlington, Vt., Oct. 27, 1831. Grad. Brown University, R.I., 1852. Lawyer and militia member in Syracuse. Lt., 12th N.Y. Inf. Lt. Col. 149th N.Y., June 8, 1863. Wd. July 3, 1863. Killed July 20, 1864 near Atlanta, Ga. When Randall wd., Capt. Nicholas Grumbach, Jr. led the regiment. b. Detroit, Mich. Tobacco business in

Syracuse, N.Y. Capt., Co. B, 149th N.Y., Sept. 2, 1862. M.O. Lt. Col., June 12, 1865. d. Syracuse, July 5, 1912, age 77 years, 5 months 4 days.

Number: 319   **Loss:** 6 k., 46 wd.

Monument: Slocum Ave. Map reference **III F-5**

Bas relief depicts color Sergt. Wm. C. Lilly (a switchman from Syracuse) mending shot through staff of flag under fire. Relief based on work done by artist Edwin Forbes.

"5 p.m. July 1,1863, occupied position near Little Round Top. 4 a.m. July 2, moved here, built these works and defended them July 2, and 3."

150th NEW YORK INFANTRY
12th Corps, 1st Div., 2nd Brig.

Other Names: "Dutchess County Regiment"

Raised: Dutchess County

Organized: Camp Dutchess, near Poughkeepsie, N.Y. M.I. Oct. 11, 1862.

Commander: Col. John Henry Ketcham. b. Dover Plains, N.Y., Dec. 21, 1832. Farmer. Elected to state senate. Col., 150th N.Y., Oct. 11, 1862. Wd. Dec. 20, 1864. M.O. Brig. Gen., Mar. 2, 1865 to take seat in U.S. Congress. d. N.Y. City, Nov. 4, 1906.

Number: 609   **Loss:** 7 k., 23 wd., 15 m.

Monument: Slocum Ave. Map reference **III F-4**

"This regiment defended these works on July 3 from 6:30 to 9 a.m. and from 10 a.m. to 12 m. and captured 200 prisoners."

Marker: Trostle House. Map reference **IV K-7**

"Charged this point on July 2, 1863, about 8 p.m. and drew off 3 abandoned guns of Bigelow's Battery."

154th NEW YORK INFANTRY
11th Corps, 2nd Div., 1st Brig.

Raised: Counties of Cattaraugus and Chautauqua.

Organized: Camp Brown, Jamestown, N.Y. M.I. Sept. 24, 1862.

Commander: Lt. Col. Daniel B. Allen. b. Otto, N.Y., April 7, 1839. 1862 practiced law in Olean, N.Y. Returned to Otto to recruit for 154th N.Y. Capt., Co. B, 154th N.Y., Sept. 26, 1862. M.O. Sept. 30, 1864. d. Buffalo, N.Y., Oct. 9, 1934.

Number: 274   **Loss:** 1 k., 21 wd., 178 m.

Monument: Coster Ave. Map reference **I H-14**

"July 1,1863. July 2 and 3, occupied position on East Cemetery Hill."

157th NEW YORK INFANTRY
11th Corps, 3rd Div., 1st Brig.

Raised: Counties of Cortland and Madison.

Organized: Camp Mitchell, Fair grounds, Hamilton, N.Y. M.I Sept. 19, 1862.

Commander: Col. Philip Perry Brown, Jr. b. Smithfield, N.Y., Oct. 8, 1823. Grad. Madison University, N.Y., 1855. Principal of the university school, 1855-1862. Col., 157th N.Y. Sept. 19, 1862. M.O. June 8, 1865. Col., Veteran Volunteers. Bakery business in St. Louis, Missouri. d. St. Louis, April 9, 1881.

Number: 431   **Loss:** 27 k., 166 wd., 114 m.

Monuments: Two monuments located along Howard Ave. Map references **I D-13, I E-10**. One was erected by the state and one by the regiment.

First monument on west Howard Ave. near its junction with Mummasburg Rd. indicates the position first held by the regt. when it was on the extreme left of the 11th Corps line on July 1, 1863.

Second monument Howard Ave. and Carlisle Rd. indicates afternoon position of regt. in brigade line.

Marker: Carlisle Rd. About 300 yards in advance of monument. Map reference **I B-13**.

Marks forward position of regt. on July 1, 1863.

163rd NEW YORK INFANTRY

See 73rd New York Infantry

1st NEW YORK ARTILLERY, BATTERY B

2nd Corps Artillery Brigade

Other Names: "Empire Battery." "Pettit's."

Raised: Onondaga County. Also men from Cook County, Ill. (originally belonging to Busteed's Chicago Battery), and New York City (belonging to one section of 14th N.Y. Battery, temporarily attached).

Organized: Elmira, N.Y. M.I. Aug. 31, 1861.

Commanders: Capt. James McKay Rorty. July 2, 1863 he was relieved from duty as 2nd Corps division ordnance officer and given command of Battery B. b. Donegal Town, Ireland, June 11, 1837. Arrived in U.S. in 1857. For a time was a book convasser in N.Y. City. Joined 69th N.Y. State militia. 2 Lt., 14th N.Y. Independent Battery, Nov. 17, 1861. Wd. Dec. 13, 1862. Killed July 3,1863. Lt. Albert S. Sheldon. b. Otsego County, N.Y. Harness maker in Baldwinsville, N.Y. Lt., Battery B, Oct. 21, 1861. Wd. July 3, 1863 and June 2, 1864. M.O. Capt., Dec. 16, 1864. d. Baldwinsville, Mar. 15, 1911, age 82 years, 5 months, 4 days. Lt. Robert Eugene Rogers took command after Sheldon. b. May 11, 1841, Newark, N.Y. Machinist in Morris, Ill. Sergt., 11th Ill. Inf. Belonged then to Busteed's Chicago Battery. Nov. 12, 1861 battery broken up and many, including Rogers, transferred to Battery B. Lt., Battery B, Sept. 10, 1862. M.O. Capt., June 18, 1865. d. Aug. 5, 1930.

Number: 4 10-pdr. Parrotts. 114 men.   **Loss:** 10 k., 16 wd.

Monument: Hancock Ave. Map reference **II E-14**

"Position held afternoon of July 3, 1863."

1st NEW YORK ARTILLERY, BATTERY C
5th Corps Artillery Brigade

Raised: Jefferson County
Organized: Elmira, N.Y. M.I. Sept. 6, 1861.
Commander: Capt. Almont Barnes. b. Turin, N.Y., April 14, 1835. Grad. National Law School, Wash., D.C. Journalist in Watertown, N.Y. Lt., Battery C, Sept. 6, 1861. M.O. Sept. 5, 1864. Served in various government posts. d. Washington, D.C., Jan. 16, 1918. Buried Arlington National Cemetery.
Number: 4 Ordnance Rifles. 88 men.   **Loss:** None
Monument: Sedgwick Ave. Map reference **V C-12**
  "Held this position from about 4:30 p.m. July 2, to 4 a.m. July 3, 1863. On the morning of July 3 transferred to the left flank of Big Round Top."
Marker: Howe Ave. Map reference **V L-16**
  Position of July 3, 3 p.m.

1st NEW YORK ARTILLERY, BATTERY D
3rd Corps Artillery Brigade

Raised: Counties of Jefferson and St. Lawrence.
Organized: Elmira, N.Y. M.I. Sept. 6, 1861.
Commander: Capt. George Bigelow Winslow. b. Adams, N.Y., June 23, 1832. Parents died when he was 16 years. Hardware business in Gouverneur, N.Y. Lt., Battery D, Oct. 21, 1861. Wd. May 5, 1864. Discharged Sept. 26, 1864. Postmaster of Gouverneur. d. Vienna, Va., Sept. 30, 1883. Number 6 12-pdr. Napoleons. 116 men. **Loss:** 10 wd., 8 m.
Monument: Wheatfield. Map reference **V B-7**
  "Held this position during the afternoon of July 2, 1863."

1st NEW YORK ARTILLERY, BATTERY E

  See 1st N.Y., Battery L

1st NEW YORK ARTILLERY, BATTERY G
Artillery Reserve, 4th Volunteer Brigade

Raised: Oswego County. Also men from Cook County, Ill. (originally belonging to Busteed's Chicago Battery), and New York City (belonging to two sections of 14th N.Y. Battery temporarily attached).
Organized: Elmira, N.Y. M.I. Sept. 24, 1861.
Commander: Capt. Nelson Ames. b. Mexico, N.Y., Nov. 24, 1836. Worked on father's Mexico farm. Lt., Battery G, Oct. 21, 1861. Wd. May 30, 1864. M.O. Oct. 15, 1864. Late in life became Mayor of Marshalltown, Iowa. d. there, Mar. 7, 1907.
Number: 6 12-pdr. Napoleons. 132 men.   **Loss:** 7 wd.
Monument: Peach Orchard. Map reference **IV K-1**

"Engaged here with 3rd Corps 3 p.m. to 5:30 p.m. July 2, 1863. July 3, on Cemetery Ridge with 1st Div., 2d Corps."

Ames—"There were no formal exercises dedicating the monument to Battery G, First Regiment, New York Light Artillery, on the battlefield of Gettysburg, July 3, 1893. But thirteen of the survivors of the battery were present, and we dedicated the noble monument in silence and in tears. No one wanted to make a speech, and none was made. Our meeting was like the meeting of a family, and formalities seemed out of place. We dedicated the monument with our tears...."

Marker: Hancock Ave. Map reference **IV J-13**

1st NEW YORK ARTILLERY, BATTERY I
11th Corps Artillery Brigade

Raised: Erie County

Organized: Buffalo, N.Y., as part of 65th N.Y. Militia. M.I. Oct. 1, 1861.

Commander: Capt. Michael Wiedrich. b. Hochorville, France, Sept. 23, 1820. Came to Buffalo in 1840. Clerk. Capt., Battery I, Aug. 20, 1861. M.O. Feb. 29, 1864. Service in 15th N.Y. Heavy Artillery. d. Buffalo, Mar. 21, 1899. Number 6 Ordnance Rifles. 141 men. **Loss:** 3 k., 10 wd.

Monument: East Cemetery Hill. Map reference **II K-3**
"July 1, 2, and 3, 1863."

1st NEW YORK ARTILLERY, BATTERY K
11th N.Y. Independent Battery Temporarily Attached
Artillery Reserve, 4th Volunteer Brigade

Raised: K-Montgomery County. 11-Counties of Albany, N.Y. and Ashtabula, Oh.

Organized: K-Albany, N.Y. M.I. Nov. 20, 1861. 11-Albany, N.Y. M.I. Jan. 6, 1862.

Commander: Capt. Robert Hughes Fitzhugh. b. Oswego, N.Y., Oct. 17, 1840. Grad. Yale College, 1861. Entered service from Oswego. Lt., Battery F, 1st N.Y. Artillery. Capt., Battery K, Oct. 14, 1862. Wd. Aug. 22, 1862 and July 30, 1864. M.O. Maj., 1st N.Y. Artillery, June 21, 1865. d. May 4, 1920 at Pittsburgh, Penn. Buried Arlington National Cemetery.

Number: 6 Ordnance Rifles (2 from 11th N.Y.). 149 men (11th N.Y.-99. Batt. K-50). **Loss:** 7 wd.

Monuments: Batteries have monuments together on Hancock Ave. Map reference **II F-11** Batt. K-"Held this position July 3 1863 and assisted in repelling Pickett's Charge."

Markers: Flank markers for Batt. K at Pleasonton and Hancock Aves. Map reference **II E-16**
Markers indicate original position of monument. Moved to present position in 1903.

1st NEW YORK ARTILLERY, BATTERY L
1st Corps Artillery Brigade

Raised: Monroe County

Organized: Recruited among the Rochester Union Grays militia unit. Organized in Elmira, N.Y. M.I. Nov. 17, 1861. 16 men of 1st New York Artillery, Battery E were temporarily attached at Gettysburg.

Commanders: Capt. Gilbert Henry Reynolds. b. N.Y. City, Dec. 18, 1832. Coachmaker in Rochester, N.Y. 2 Lt., Battery L, Nov. 12, 1861. Wd. July 1, 1863. Discharged for disability, May 3, 1864. d. Rochester, April 6, 1913. Lt. George Breck took command from the wounded Reynolds at Gettysburg. b. Newport, New Hampshire, Aug. 18, 1833. Druggist in Rochester, N.Y. Member of militia. Pvt., Battery L, Sept. 25, 1861. M.O. Capt., June 17, 1865. d. Hubbardstown, Mass., Oct. 5, 1925. Buried Arlington National Cemetery.

Number: 6 Ordnance Rifles. 141 men.   **Loss:** 1 gun captured July 1, 1863. 1 k., 15 wd., 1 m.

Monuments: Reynolds Ave. and East Cemetery Hill.

Map references I H-4, II L-4

East Cemetery Hill-"These works were built and held by Battery 'L', Lieutenant George Breck commanding, against assaults of infantry and artillery during the second and third days of July, 1863." Breck wrote on the morning of July 2, "we are now in position near Gettysburg Cemetery[,] a very high and commanding position. Both armies are concentrating all their forces in this vicinity, and I dare say, the greatest battle of the war will be fought at this place."

1st NEW YORK ARTILLERY, BATTERY M
12th Corps Artillery Brigade

Other Names: "Cothran's"

Raised: Niagara County

Organized: Lockport, N.Y. M.I. Oct. 14, 1861.

Commander: Lt. Charles E. Winegar. b. Canaan, N.Y., Sept. 8, 1832. Paper manufacturer in Shelby, N.Y. Lt., Battery M, Oct. 14, 1861. Wd. May 25, 1862. M.O. Capt., June 23, 1865. In 1884, Winegar lived in Brooklyn, N.Y. Buried Cypress Hills National Cemetery.

Number: 4 10-pdr. Parrotts. 96 men.   **Loss:** None

Monument: Powers Hill. Map reference III E-17

"Held this position July 2d-3d, 1863."

1st NEW YORK INDEPENDENT BATTERY
6th Corps Artillery Brigade

Raised: Cayuga County

Organized: Auburn, N.Y. M.I. Nov. 23, 1861.

Commander: Capt. Andrew Cowan. b. Ayrshire, Scotland, Sept. 29, 1841. Start of the war lived in Auburn, N.Y. Student at Madison University, N.Y.

Served in 19th N.Y. Infantry and 3rd N.Y. Artillery. Lt., 1st Indpt. Battery, Nov. 23, 1861. Wd. Sept. 13, 1864. M.O. June 23, 1865. Established the firm of Andrew Cowan & Co. in Louisville, Kentucky. d. Louisville, Aug. 23, 1919.

Number: 6 Ordnance Rifles. 113 men.   **Loss:** 4 k., 8 wd.

Monument: Hancock Ave. Map reference **II F-13**

"During the cannonade preceding Longstreet's assault, the battery was engaged a short distance farther to the left, but by order of General Webb, it moved at a gallop to this position, which Battery B, 1st R.I. artillery had occupied. Skirmishing had just commenced. The Confederate lines were advancing and continued their charge in the most splendid manner up to our position. The artillery fire was continuous and did much execution. Our last charge, double canister, was fired when some of the enemy were over the defenses and within ten yards of our guns."

The final moments of the charge, as described by Cowan, served as the basis for the monument's bas relief. "Young McElroy had thrust the canister into the muzzle and fell dead in front of the wheel, with three rifle balls in his face. Gates rammed the charge home and springing back, fell shot through both legs. Bassenden sprang forward to seize the sponge staff. Stears (W.A.) pricked the cartridge and was ready to fire at the word. Little "Aleck" McKenzie, the corporal, was running down the sight, to meet the enemy just crossing the demolished stone wall ten yards away, and as he signaled 'Fire!' he fell across the trail of the gun, wounded. The leading horses of the gun were both down. The nigh horse of the swing team was also shot, but the wheel-team stood unhurt; and mounted there was the driver, that 'wild Irishman', 'Mike' Smith, crazy with excitement."

3rd NEW YORK INDEPENDENT BATTERY
6th Corps Artillery Brigade

Raised: New York City

Organized: Camp Anderson, at the Battery, as part of the 2nd N.Y. State Militia (82 N.Y.). M.I. June 17, 1861.

Commander: Capt. William A. Harn. b. Philadelphia, Penn., 1834. Served in 1st U.S. Artillery, 1854-Sept. 5, 1862. Lt., Battery G, 1st N.Y. Artillery. Capt., 3rd Indpt. Battery, April 13, 1863. M.O. June 24, 1865. Late in life became light house keeper. d. St. Augustine, Florida, May 31, 1889.

Number: 6 10-pdr. Parrotts. 119 men.   **Loss:** None

Monument: Taneytown Road. Map reference **II I-8**

"Forced march 36 miles, second position."

4th NEW YORK INDEPENDENT BATTERY
3rd Corps Artillery Brigade

Raised: New York City

Organized: Staten Island, N.Y. as part of Serrell's 1st N.Y. Engineers. M.I. Oct. 24, 1861.

Commander: Capt. James Edward Smith. b. Schenectady, N.Y., Aug. 8, 1832. Livestock commission merchant in N.Y. City. Member of militia battery. Capt., 4th Indpt. Battery, Oct. 15, 1861. M.O. Dec. 21, 1863. d. Washington, D.C., April 18, 1893. Buried Arlington National Cemetery.
Number: 6 10-pdr. Parrotts. 135 men.   **Loss:** 2 k., 10 wd., 1m.
Monument: Sickles Ave. Map reference **V E-7**
"At the time of the assault by Hood's Division of the Confederate army this battery supported by the Fourth Maine Infantry, formed the extreme left of the Third Corps line. Three guns of the two sections in action on this crest were captured by the Confederates. The third section was in position to the right and rear and continued the action until nearly 6 p.m."
Marker: Crawford Ave. (3rd section). Map reference **V D-9**

5th NEW YORK INDEPENDENT BATTERY
Artillery Reserve, 2nd Volunteer Brigade

Other Names: "First Excelsior Light Artillery"
Raised: New York City. Counties of Monroe and Kings.
Organized: Brooklyn, N.Y. as part of General Sickles' Excelsior Brigade. M.I. Nov. 8, 1861.
Commander: Capt. Elijah D. Taft. b. Mamaroneck, N.Y., April 28, 1819. Carpenter and joiner in Brooklyn, N.Y. Militia. Capt., 1st Excelsior (5th Indpt. Battery), Aug. 21, 1861. M.O. July 16, 1865. d. Freeport, N.Y., Mar. 1, 1915.
Number: 6 20-pdr. Parrotts. 146 men.   **Loss:** 1 k., 2 wd.
Monument: National Cemetery. Map reference II J-4
"This battery held this position from 5 p.m. July 2 to 5, 1863."
Markers: Two tablets on Baltimore Pike and in Evergreen Cemetery Map references **II J-4, II K-4**
Evergreen Cemetery-"July 2. Arrived and halted in park about 10:30 a.m. Moved to the cemetery at 3:30 p.m. and engaged from 4 p.m. until dark. Four guns south of and facing Baltimore Pike firing on a Confederate battery on Benner's Hill. Two guns firing westwardly.
July 3. Engaged at intervals in same position until 4 p.m. One gun on Baltimore Pike having burst, the other three relieved the section firing westwardly. Remained in this position until the close of the battle."

6th NEW YORK INDEPENDENT BATTERY
Cavalry Corps, Reserve Artillery Brigade

Raised: New York City. Union County, N.J.
Organized: N.Y. City as part of 9th N.Y. State Militia (83rd N.Y. Inf.). M.I. June 15, 1861.
Commander: Capt. Joseph William Martin. b. New York, Nov. 3, 1838. Grad. Rutgers College, N.J., 1855. Clerk in N.Y. City living in Rahway, N.J. 2 Lt., 6th Independent Battery, June 15, 1861. M.O. Feb. 16, 1865. d. Rahway, N.J., Nov. 13, 1908.

Number: 6 Ordnance Rifles. 130 men. **Loss:** 1 wd.
Monument: Taneytown Road. Map reference **II H-9**
"Occupied this position July 3, 1863."

10th NEW YORK INDEPENDENT BATTERY

No distinct organization at Gettysburg. By special order 151, June 1863, the battery was broken up. Men temporarily distributed to Battery G, 1st R.I. (32 men), Battery C, 1st R.I. (20 men), 5th Mass. Battery (19 men), and 5th N.Y. Independent Battery (27 men).

11th NEW YORK INDEPENDENT BATTERY

See 1st New York Artillery, Battery K

13th NEW YORK INDEPENDENT BATTERY
11th Corps Artillery Brigade

Other Names: "Baker's Brigade Artillery"
Raised: New York City
Organized: N.Y. City as part of "Philadelphia Brigade." M.I. Oct. 15, 1861.
Commander: Lt. William Wheeler. b. N.Y. City, Aug. 14, 1836. Grad. Yale College, 1855. Attnd. Harvard University. Lawyer in N.Y. City. Militia service. Lt., 13th N.Y. Indpt. Battery, Oct. 15, 1861. Capt. Commission was later determined to date to May 26, 1863. Killed June 22, 1864 near Marietta, Ga. Wheeler said of Gettysburg, "somehow or other I felt a joyous exaltation, a perfect indifference to circumstances through the whole of that three day's fight, and have seldom enjoyed three days more in my life ...."
Number: 4 Ordnance Rifles. 118 men. **Loss:** 8 wd., 3 m.
Monument: Howard Ave. Map reference **I D-12**
"July 1,1863 engaged here. July 2, on Cemetery Hill. July 3, at repulse of Pickett's charge."

14th NEW YORK INDEPENDENT BATTERY

No distinct organization at Gettysburg. Two sections serving temporarily with 1st N.Y. Artillery, Battery G. One section serving temporarily with 1st N.Y. Artillery, Battery B.

15th NEW YORK INDEPENDENT BATTERY
Artillery Reserve, 1st Volunteer Brigade

Raised: New York City
Organized: Fort Schuyler, Throgs Neck, N.Y. City, as part of the "Irish Brigade." M.I. Dec. 9, 1861.
Commander: Capt. Patrick Hart. b. County Leitrim, Ireland, 1827 or 1828. Regular service in the United States (artillery, ordnance, and Marines), 1845-

1862. Capt., 15th N.Y. Indpt. Battery, Mar. 5, 1863. Wd. July 3, 1863 and June 8, 1864. Transferred to 32nd N.Y. Artillery. M.O. July 12, 1865. After brief period of Regular service, became superintendent of two national cemeteries. d. Richmond, Va., Jan. 22, 1892.

Number: 4 12-pdr. Napoleons. 99 men. **Loss:** 3 k., 13 wd.

Monument: Wheatfield Rd., opposite Peach Orchard. Map reference **IV L-2** "July 2, 1863."

Marker: Hancock Ave. **IV H-13**

"July 2. Engaged in the Peach Orchard. Retired about dark and reported to Brig. General R.O. Tyler Artillery Reserve.

July 3. Ordered early to the front and took position in the battalion on the left of Battery E, 5th Massachusetts. Directed by Maj. General Hancock to open on the Confederate batteries with solid shot and shell. Upon the advance of Confederate infantry, fired shell and shrapnel and canister when the line was within 500 yards. A second line advancing was met with double canister which dispersed it. The fire of the battery was then directed against the artillery on the Confederate right and several caissons and limbers were exploded by the shells.

July 4. Remained in this position until noon."

ONEIDA INDEPENDENT COMPANY, NEW YORK CAVALRY
Army Headquarters

Raised: Madison County

Organized: Oneida, N.Y. M.I. Sept. 4, 1861.

Commander: Capt. Daniel P. Mann. b. Smithtown, N.Y., 1811 or 1812. Cooper before enlisting in Regular service, 1833-1836. Quartermaster Dept. during Mexican War. Enlisted from residence in Canastota, N.Y. Capt., Oneida Company, Sept. 4, 1861. M.O. Dec. 10, 1864.

Number: 49   **Loss:** None

Monument: Taneytown Road. Map reference **II H-10**

"General Meade's escort and headquarters orderlies and couriers."

2nd NEW YORK CAVALRY

Detached with its brigade in Maryland during the Battle of Gettysburg.

4th NEW YORK CAVALRY

Detached with its brigade in Maryland during the Battle of Gettysburg.

5th NEW YORK CAVALRY
11 Companies
Cavalry Corps, 3rd Div., 1st Brig.

Other Names: "First Ira Harris Guard"

Raised: New York City. Counties of Essex and Tioga.

Organized: Camp Scott, Staten Island, N.Y. M.I. Aug. - Oct., 1861. Company I was in Washington, D.C. during the battle.

Commander: Maj. John Hammond. b. Crown Point, N.Y., Aug. 17, 1827. Attnd. Rensselaer Polytechnic Institute, N.Y. California pioneer in 1849. Iron manufacturer in Crown Point. Capt., Co. H 5th N.Y. Cavalry, Oct. 18, 1861. Wd. June 1, 1864 and Sept. 13, 1863. M.O. Sept. 3, 1864. Elected to U.S. Congress, 1879-1883. d. Crown Point, May 28, 1889.

Number: 468 **Loss:** 1 k., 1 wd., 4 m.

Monument: Hill southwest of Big Round Top. Map reference **V K-4**
"July 3,1863, this regiment under command of Maj. John Hammond here supported Battery E, 4th U.S. Horse Artillery."

6th NEW YORK CAVALRY
Cavalry Corps, 1st Div., 2nd Brig.
Company D and K, 2nd Corps Headquarters
Company L, 1st Cavalry Div., 2nd Brig., Provost Guard
Companies F & H at Yorktown, Va.
Company A, 3rd Corps Headquarters

Other Names: "Second Ira Harris Guard"

Raised: New York City. Counties of Columbia, Steuben and St. Lawrence.

Organized: Camp Scott and Camp Herndon, Staten Island, N.Y. M.I. Sept. - Dec., 1861.

Commander: Maj. William Elliott Beardsley. b. near Norwalk, Conn., Mar. 11, 1826. Tailor in N.Y. City. Corpl., 71st N.Y. State Militia. Capt., Co. D (E), 6th N.Y. Cavalry, Oct. 12, 1861. M.O. Oct. 21, 1864. Maj., 26th N.Y. Cavalry, Mar. 15, 1865. M.O. Lt. Col., July 7, 1865. d. Akron, Oh., Dec. 25, 1884.

Number: 407 **Loss:** 1 k., 3 wd., 8 m.

Monument: Buford Ave. Map reference **I C-5**
"Arrived June 30, 1863. July 1 skirmished dismounted, on this line until arrival of 1st Corps and the rest of the day on right of the York road, then retired to Cemetery Hill, one squadron being among the last Union troops in Gettysburg on that day. Bivouacked in Peach Orchard that night, and engaged enemy's skirmishers on the morning of July 2, until relieved by troops of the 3d Corps; then moved to Taneytown and on the third to Westminister, from which place moved with the Division (Buford's) in pursuit of the enemy."

8th NEW YORK CAVALRY
Cavalry Corps, 1st Div., 1st Brig.

Other Names: "Rochester Regiment"

Raised: Counties of Monroe, Niagara and Chenango.

Organized: Rochester fair grounds, Rochester, N.Y. M.I. Nov. 23 and 28, 1861.

Commander: Lt. Col. William Lester Markell. b. Manheim Centre, N.Y., Jan. 15, 1836. Manufacturer and wholesaler in Rochester. Militia service. Maj.,

8th N.Y. Cavalry, Nov. 28, 1861. Wd. July 10, 1863. M.O. Feb. 27, 1864. d. Syracuse, N.Y., Feb. 13, 1916.

Number: 623   **Loss:** 2 k., 22 wd., 16 m.

Monument: Reynolds Ave. Map reference I I-4

"Pickets of this regiment were attacked about 5 a.m., July 1, 1863, by the advance skirmishers of Heth's Confederate Division; the regiment engaged the enemy west of Seminary Ridge, with the brigade stubbornly contesting the ground against great odds until about 10:30 a.m., when it was relieved by the advance regiments of the 1st Corps."

9th NEW YORK CAVALRY
Cavalry Corps, 1st Div., 2nd Brig.
Companies D&L, 12th Corps Headquarters

Other Names: "Westfield Cavalry"

Raised: Counties of Chautaugua, Cattaraugus, Wyoming and St. Lawrence.

Organized: Camp Seward, fair grounds, Westfield, N.Y. M.I. Oct. 1-3, 1861.

Commander: Col. William Sackett. b. Seneca Falls, N.Y., April 16, 1839. Lawyer in Chicago, Ill. at the start of the war. Sergt., 19th Ill. Infantry. Maj., 9th N.Y. Cavalry, Nov. 20, 1861. Mort. wd. June 11, 1864 at Trevilian Station, Va. d. June 14, 1864.

Number: 395   **Loss:** 2 k., 2 wd., 7m.

Monument: Buford Ave. Map reference I B-6

"Position 8 a.m. July 1, 1863. Picket on Chambersburg Road, fired on at 5 a.m."

10th NEW YORK CAVALRY
Cavalry Corps, 2nd Div., 3rd Brig.

Other Names: "Porter Guard"

Raised: Counties of Erie, Chemung, Chenango, Cortland, Fulton, and Onondaga.

Organized: Elmira, N.Y. M.I. Dec. 23, 1861.

Commander: Maj. Mathew Henry Avery. b. Middletown Springs, Vt., Mar. 27, 1835. Book and stationary business in Syracuse, N.Y. Capt., Co. A, 10th N.Y. Cav., Sept. 27, 1861. Wd. Dec. 1, 1864. Transferred to 1st N.Y. Provisional Cavalry. M.O. Col., Aug. 4, 1865. Pioneer in oil business. d. Geneva, N.Y., Sept. 1, 1881.

Number: 392   **Loss:** 2 k., 4 wd., 3 m.

Monument: Hanover Road. See key map.

"July 2nd 1863. 3 to 8 p.m."

4th OHIO INFANTRY
2nd Corps, 3rd Div., 1st Brig.

Raised: Counties of Knox, Delaware, Hardin, Marion, Wayne and Stark.

Organized: Camp Dennison near Cincinnati, Oh. M.I. June 5, 1861.

Commander: Lt. Col. Leonard Willard Carpenter. b. Indiana, Penn., Jan. 26, 1834. Medical student in Mt. Vernon, Oh. Lt., Co. A, 4th Oh., April 27, 1861 (3 mos.). M.O. of 3 yrs. regiment, June 21, 1864. d. Seattle, Washington, Feb. 18, 1908. Buried in Gettysburg National Cemetery in 1910.

Number: 229   **Loss:** 9 k., 17 wd., 5 m.

Monument: East Cemetery Hill. Map reference II K-4

"On the evening of July 2, 1863, Carroll's Brigade was sent from its position with the 2nd Corps to re-enforce this portion of the line, and this monument marks the position where, as part of that brigade, the 4th Ohio Infantry at that time participated in repelling an attack of the enemy."

Marker: Emmitsburg Road, companies G&I. Map reference II E-8

"At 3 p.m., July 2, 1863, while the regiment was lying on Cemetery Ridge, Companies G and I, Fourth Ohio Infantry, detached under Captain Peter Grubb of Company G, advanced to this position where, with severe loss, they engaged the enemy during the remainder of the day. Late in the evening they were withdrawn to the regiment on East Cemetery Hill."

5th OHIO INFANTRY
12th Corps, 2nd Div., 1st Brig.

Raised: Hamilton County.

Organized: Camp Dennison, near Cincinnati, Oh. M.I. June 21, 1861.

Commander: Col. John Halliday Patrick. b. Edinburgh, Scotland, Mar. 11, 1820. Came to Cincinnati in 1848. Tailor in Cincinnati. Militia service. Lt. Col., 5th Oh., June 21, 1861. Mortally wounded May 25, 1864 at New Hope Church, Ga.

Number: 315   **Loss:** 2 k., 16 wd.

Monument: Geary Ave. Map reference III E-7

"Arriving in position at 5 p.m., July 1, was detached and held extreme left of line on north side of Little Round Top. Morning of July 2 moved to Culp's Hill, and at evening moved as far as Rock Creek to re-enforce the left. Returned to Culp's Hill during the night and on morning of July 3 was engaged where this monument stands until 11 a.m. in repulsing the enemy and retaking the Union works."

Marker: Small stone on Sykes Ave., north slope of Little Round Top. Map reference V D-12

7th OHIO INFANTRY
12th Corps, 2nd Div., 1st Brig.

Raised: Counties of Cuyahoga, Portage, Lorain, Lake, Huron, Trumbull and Mahoning.

Organized: Camp Dennison, near Cincinnati, Oh. M.I. June 21, 1861.
Commander: Col. William R. Creighton. b. Pittsburgh, Penn., June 1837. Printer in Cleveland, Ohio. Member of militia. Capt., Co. A, 7th Oh. (3 mos.), April 22, 1861. Wd. Aug. 9, 1862. Killed in fighting at Ringgold, Ga., Nov. 27, 1863.
Number: 293 **Loss:** 1 k., 17 wd.
Monument: Slocum Ave. Map reference III F-6

"Arrived near Little Round Top, evening of July 1. On July 2, held positions on Culp's Hill from morning until 6 p.m., then moved with Brigade to support the left. Returned at midnight to Culp's Hill, and remained there until close of the battle."

8th OHIO INFANTRY
2nd Corps, 3rd Div., 1st Brig.

Raised: Counties of Sandusky, Seneca, Cuyahoga, Crawford, Huron, Erie, Lorain and Medina.
Organized: Camp Dennison, near Cincinnati, Oh. M.I. June 24, 1861.
Commander: Lt. Col. Franklin Sawyer. b. Auburn, Oh., July 13, 1825. Attnd. Denison University, Granville, Oh. Lawyer, militia officer and Prosecuting Attorney for Huron County. Residence in Norwalk, Ohio. Capt. Co. D, 8th Oh. (3 mos.), April 29, 1861. Wounded July 3, 1863 and May 12, 1864. M.O. July 13, 1864. d. Norwalk, Ohio. Aug. 22, 1892.
Number: 209 **Loss:** 18 k., 83 wd., 1 m.
Monument: Emmitsburg Road. Map reference II D-9

"The 8th Ohio Infantry under Lieut. Col. Franklin Sawyer, took this position at 4 p.m. July 2, after a brief skirmish and held it July 2 and 3 during Longstreet's assault. July 3, the regiment advanced and by left wheel attacked the enemy in flank, capturing three flags and numerous prisoners."

25th OHIO INFANTRY
9 Companies
11th Corps, 1st Div., 2nd Brig.

Raised: Counties of Belmont, Monroe, Sandusky, Morgan, Noble and Lucas.
Organized: Camp Chase, Columbus, Oh. M.I. June 28, 1861. Co. D became 12th Ohio Battery in 1862.
Commanders: Lt. Col. Jeremiah Williams. b. near New Martinsville, Virginia (West Virginia), Mar. 23, 1832. Editor in Woodsfield, Oh. Capt., Co. C, 25th Oh., June 10, 1861. Captured July 1, 1863. M.O. June 21, 1864. d. Brooklyn, N.Y., May 11, 1915. Buried Arlington National Cemetery. When Williams captured, Capt. Nathaniel James Manning took command of the regiment b. Randolph County, Illinois, Feb. 14, 1837. Attnd. University of Michigan. Lawyer in Woodsfield, Oh. Sergt., Co. C, 25th Oh., June 10, 1861. Wd. July 1, 1863. M.O. June 26, 1864. Maj., 191st Oh. M.O. Aug. 27, 1865. d. Barnesville, Oh., Mar. 11, 1874. For a time on July 2, 1863, 2 Lt. William Maloney led the regiment. b. Leesville, Oh., June 5, 1837. Regular army service, 1855-1860.

Farmer in Leesville. Sergt., Co. F, 25th Oh., June 13, 1861. M.O. July 26, 1864. d. Harrah, Oklahoma, Nov. 14, 1912. Final commander of unit at Gettysburg was Lt. Israel White. b. Belmont County, Oh. around 1821. Laborer in St. Clairsville, Oh. Sergt., Co. A, 25th Oh., June 5, 1861. Wd. June 8, 1862, May 2, 1863 and Nov. 30, 1864. M.O. Capt., June 18, 1866. Ran a saloon in Columbia, S.C. until 1867, when, starting for his home in Ohio, he disappeared, never to be heard from again.

Number: 280 **Loss:** 9 k., 100 wd., 75 m.

Monuments: 25th Oh. and 75th Oh. combined. Howard Ave. and Wainwright Ave. Map references **I C-14, II K-2**
See entry for 75th Oh. for inscription.

29th OHIO INFANTRY
12th Corps, 2nd Div., 1st Brig.

Other Names: "Giddings Regiment." Named in honor of Congressman Joshua R. Giddings of Ohio. See also 14th U.S. Infantry.

Raised: Counties of Ashtabula, Summit and Medina.

Organized: Camp Giddings, near Jefferson, Oh. M.I. Aug. 26, 1861.

Commanders: Capt. Wilber F. Stevens. b. Ohio, 1839 or 1840. Volunteer service during Utah expedition,1857-1858. Student in Pierpont, Oh. Capt., Co. B, 29th Oh., Aug. 19, 1861. Wd. July 3, 1863 and May 25, 1864. M.O. Sept. 9, 1864. d. Tacoma, Washington, May 12, 1894. When Stevens wounded, Capt. Edward Hayes took command. b. Hartford, Oh., Sept. 30, 1829. Farmer in Hartford at the start of the war. Capt., Co. C, 29th Oh., Sept. 7, 1861. Wd. May 8, 1864. M.O. Lt. Col., Nov. 4, 1864. d. Park Hotel in Warren, Ohio, Aug. 18, 1899.

Number: 332 **Loss:** 7 k., 31 wd.

Monument: Slocum Ave. Map reference **III F-5**
"The 29th Ohio Infantry, commanded by Captain Edward Hayes, J. B. Storer Adjutant-occupied several positions in this vicinity, both in the intrenchments and in reserve, July 2 and 3,1863." Stevens' name is missing from monument perhaps because Adjt. Storer thought him a coward at the Battle of Cedar Mountain, Va., and receiving only a superficial wound at Gettysburg.

55th OHIO INFANTRY
11th Corps, 2nd Div., 2nd Brig.

Raised: Counties of Huron, Erie, Sandusky, Seneca and Wyandot.

Organized: Camp McClellan near Norwalk, Oh. M.I. Jan. 25, 1862.

Commander: Col. Charles B. Gambee. b. on a farm in Seneca County, N.Y., April 5, 1827. Dry goods business in Bellevue, Oh. Capt., Co. A, 55th Oh., Sept. 30, 1861. Killed at Resaca, Ga., May 15, 1864.

Number: 375 **Loss:** 6 k., 31 wd., 12 m.

Monument: Emmitsburg Road. Map reference **II H-3**

"Arrived at 2:20 p.m., July 1, in this position throughout the battle with severe loss. Its skirmishers drove back those of the enemy and seized a barn between the lines, where twelve of its men were surrounded and captured by the enemy's main line."

61st OHIO INFANTRY
11th Corps, 3rd Div., 1st Brig.

Raised: Counties of Hamilton, Belmont and Pickaway.
Organized: Camp Chase, Columbus, Oh. M.I. April 23, 1862.
Commander: Col. Stephen Joseph McGroarty. b. Mt. Charles, County Donegal, Ireland, 1830. Came to U.S. when 3 yrs. old. Attnd. St. Francis Xavier College, Oh. Lawyer in Cincinnati, Oh. Capt., 10th Ohio Infantry. M.O. April 23, 1862 to accept position as Lt. Col. in 61st Oh. Wd. July 20, 1864. Left arm amputated. M.O. July 24, 1865. d. College Hill, Oh., Jan. 2, 1870.
Number: 306 **Loss:** 6 k., 36 wd., 12 m.
Monument: Howard Ave. Map reference I E-12
"The 61st Ohio Infantry, on arriving from Emmittsburg [sic] about one o'clock p.m. July 1, 1863, was deployed as a skirmish line in advance of its brigade, and moved towards Oak Hill. Later it supported a section of Dilger's Battery, and engaged the enemy on this ground. After an obstinate contest it withdrew with the 11th Corps to Cemetery Hill. On the evening of July 2nd it moved to the assistance of the 12th Corps on Culp's Hill, and returning lay on Cemetery Hill during the remainder of the battle."
Marker: National Cemetery. Map reference II I-4

66th OHIO INFANTRY
12th Corps, 2nd Div., 1st Brig.

Raised: Counties of Champaign, Delaware, Union and Logan.
Organized: Camp McArthur, near Urbana, Oh. M.I. Dec. 1861.
Commander: Col. Eugene Powell. b. Delaware, Oh., Nov. 16, 1834. Attend. Ohio Wesleyan University, Oh. Machinist in Meadville, Penn., and Delaware, Oh. Capt., 4th Oh. Inf. (3 mos.) Maj., 66th Oh., Dec. 6, 1861. Wd. Sept. 17, 1862. Served also in 193rd Oh. Infantry. M.O. Aug. 4, 1865. Elected to state legislature. d. Marion Township, Franklin County, Oh., Mar. 17, 1907.
Number: 316 **Loss:** 17 wd.
Monument: Slocum Ave. Map reference III F-3
"The 66th Ohio Infantry arrived in position just north of Little Round Top at 5 p.m. July 1. Morning of July 2 moved to Culp's Hill and intrenched. At daybreak July 3 advanced over the Union breastworks, and with right here and left at tablet below, opened an enfilading fire upon the enemy."
Marker: Near Monument. Map reference III F-3
Marker locates position of regiment and indicates the spot where Maj. Joshua G. Palmer, a dentist from Urbana, Oh., was mortally wounded.

73rd OHIO INFANTRY
11th Corps, 2nd Div., 2nd Brig.

Raised: Counties of Ross, Pike, Highland, Pickaway and Athens.
Organized: Camp Logan, near Chillicothe, Oh. M.I. Dec. 30, 1861.
Commander: Lt. Col. Richard Long, Jr. b. Ohio, 1837. Father long time resident of Chillicothe, Oh. In 1860, Richard, Jr. was living in Groveport, Oh. Sergt., 26th Ohio Infantry. 2 Lt., 73rd Oh., Oct. 4, 1861. M.O. June 27, 1864. Inventor and manufacturer of Long's Truss Rail Joint. Run over by a train and killed at Pittsburgh, Pa., April 4, 1889. Buried Grandview Cemetery in Chillicothe.
Number: 450 **Loss:** 21 k., 120 wd., 4 m.
Monument: Taneytown Road. Map reference II H-4
"July 1, 2, 3, 1863."

75th OHIO INFANTRY
11th Corps, 1st Div., 2nd Brig.

Raised: Counties of Hamilton, Preble, Warren, Athens and Vinton.
Organized: Camp John McLean, near Cincinnati, Oh. M.I. Nov. 7, 1861.
Commanders: Col. Andrew Lintner Harris. b. Wilford Township, Butler County, Oh., Nov. 17, 1835. Grad. Miami University, Oh., 1860. Law student in Eaton, Oh., when the war began. 2 Lt., 20th Oh. Infantry (3 mos.). 2 Lt., Co. C, 75th Oh., Oct. 3, 1861. Wd. May 8, 1862. M.O. Jan. 17, 1865. Member of state legislature and Governor of Ohio, 1906-1909. d. Eaton, Oh., Sept. 13, 1915. When Harris took command of the brigade on July 1, Capt. George Benson Fox led the regiment. b. Cincinnati, Oh., June 5, 1843. Bookkeeper in Lockland, Oh. Corpl., 11th Indiana Infantry. 2 Lt., Co. B, 75th Oh., Nov. 7, 1861. M.O. Maj., Mar. 19, 1865. d. Wyoming, Oh., Sept. 20, 1924.
Number: 285 **Loss:** 16 k., 74 wd., 96 m.
Monuments: Howard and Wainwright Aves. 25th and 75th combined. Map references I C-14, II K-2
Howard Ave—"This monument marks the left flank of the 25th and the right flank of the 75th Ohio Infantry, July 1, 1863. Arriving at Gettysburg from Emmittsburg [sic] July 1, 1863, the 25th and 75th Ohio Infantry advanced beyond the town, and, under a heavy cannonade, took position here, supporting Battery G, 4th U.S. Artillery. During July 2 and 3, they held an advanced line on East Cemetery Hill, and early July 4 led the advance into the town."
Wainwright Ave—"After a severe battle in the open fields beyond Gettysburg on July 1,1863, the 11th Corps withdrew to Cemetery Hill, and at dark on July 2 this position was held by the 25th and 75th Ohio Infantry when Early's Confederate Division assaulted this hill and broke the Union line to the right, but was repulsed after a desperate hand-to-hand conflict."

82nd OHIO INFANTRY
11th Corps, 3rd Div., 2nd Brig.

Raised: Counties of Hardin, Marion, Logan, Union and Ashland.

Organized: Camp Simon Kenton, Kenton, Oh. M.I. Dec. 31, 1861.

Commanders: Col. James Sidney Robinson. b. near Mansfield, Oh. Oct. 14, 1827. Editor and newspaper publisher in Kenton, Oh. Lt., 4th Oh. Inf., Dec. 31, 1861 (3 mos.). Maj., 82nd Oh., Dec. 31, 1861. Wd. July 1, 1863. M.O. Brig. Gen. Aug. 31, 1865. Served two terms in Congress. Secretary of State of Ohio, 1885-1889. d. Kenton, Oh., Jan. 14, 1892. When Robinson wd., Lt. Col. David Thomson took command. b. near Marion, Oh., April 19, 1825. Banker in Kenton, Oh. 2 Lt., Co. A, 82nd Oh., Oct. 4, 1861. Wd. May 25, 1864. M.O. May 15, 1865. d. Kenton, Oh., Feb. 2, 1893.

Number: 384 **Loss:** 17 k., 84 wd., 79 m.

Monument: Howard Ave. Map reference **I D-13**

"The 82nd Ohio Infantry, arriving from Emmittsburg [sic] at noon, July 1, 1863, moved rapidly to the support of Dilger's Battery near the Carlisle Road. At 3 p.m. changed front to the right and advanced to a position 125 yards in front of this monument, where, exposed both front and flank to a severe fire, it engaged the enemy then approaching from York. After an obstinate struggle, the regiment, being outflanked on both sides, withdrew to Cemetery Hill, where it remained until the close of the battle."

Markers: National Cemetery. Map reference **II J-3**

107th OHIO INFANTRY
11th Corps, 1st Div., 2nd Brig.

Raised: Counties of Stark, Cuyahoga, Wayne, Lorain, Erie and Defiance.

Organized: Camp Cleveland, Railroad Street, Cleveland, Ohio. M.I. Aug. 25, 1862.

Commander: Capt. John Michael Lutz. b. Bavaria, Germany, 1829. Shoemaker in Cleveland, Oh. 2 Lt., Co. E, 107th Oh., Sept. 2, 1862. M.O. Aug. 20, 1864. d. Cleveland, April 18, 1891.

Number: 480 **Loss:** 23 k., 111 wd., 77 m.

Monument: Howard Ave. Map reference **I C-14**

"The 107th Ohio Infantry left Emmittsburg [sic] at 8 a.m. and reached Gettysburg at 1 p.m. July 1. Engaged the enemy with the brigade, losing heavily. Subsequently fell back to East Cemetery Hill, and there formed in front of Wiedrich's Battery. Evening of July 2, participated in repulsing the attack of Hays' Louisiana Brigade, Adjutant P. F. Young capturing the colors of the 8th Louisiana 'Tigers.' July 3 remained on East Cemetery Hill, exposed to fire of sharpshooters and artillery. Early July 4, made a sortie to the town."

Two markers: East Cemetery Hill. Map reference **II K-1**

1st OHIO ARTILLERY, BATTERY H
Artillery Reserve, 3rd Volunteer Brigade

Other Names: "Huntington's"
Raised: Counties of Lucas and Washington.
Organized: Camp Dennison, near Cincinnati, Oh. M.I. Nov. 7, 1861. In June 1863, 24 men of 1st Penn. Artillery, Battery F, joined Battery H.
Commander: Lt. George W. Norton. b. New York, 1818. Farmer living near Toledo, Oh. 2 Lt., Battery H., Oct. 7, 1861. M.O. Capt., Mar. 21, 1864. Lived Lucas County. d. Jan. 11, 1906.
Number: 6 Ordnance Rifles. 123 men. **Loss:** 2 k., 5 m.
Monument: National Cemetery. Map reference II I-6
"July 2nd and 3rd, 1863."

1st OHIO ARTILLERY, BATTERY I
11th Corps Artillery Brigade

Raised: Hamilton County
Organized: Gamp Dennison, near Cincinnati, Oh. M.I. Dec. 3, 1861.
Commander: Capt. Hubert Dilger. D. Baden, Germany, Mar. 5, 1836. Came to U.S. at start of Civil War. Entered military service in April 1862 as a captain of artillery under General John C. Fremont. He claimed his occupation was a German artillery officer. Capt., Battery I, Nov. 17, 1862. M.O. July 24, 1865. In 1893 he was awarded Medal of Honor for actions in Battle of Chancellorsville. Late in life he settled near Front Royal, Va. d. May 14, 1911.
Number: 6 12-pdr. Napoleons. 137 men. **Loss:** 13 wd.
Monument: Howard Ave. Map reference I D-13
"This battery, Captain Hubert Dilger commanding, marched with the 11th Corps from Emmittsburg [sic] to Gettysburg July 1. At once upon arriving it advanced rapidly on the Carlisle Road, and having taken position near this spot, immediately engaged the enemy. Reinforced by Wheeler's New York Battery, Captain Dilger advanced twice from this position. Retired with the 11th Corps, but halted and again engaged the enemy before crossing the bridge into the town. During the remainder of the battle the battery held the extreme right of Major Osborn's line on Cemetery Hill."
Markers: Howard Ave. and National Cemetery. Map references I D-12, II J-4
Howard Ave—"Battery I, 1 O.L.A. 2 guns were posted 100 yds. in rear of this tablet."

1st OHIO ARTILLERY, BATTERY K
11th Corps Artillery Brigade

Raised: Counties of Washington and Cuyahoga.
Organized: Originally to be formed in Virginia (West Virginia), but due to lack of recruits, the battery was offered to, and accepted by, the state of Ohio. Organized at Camp Dennison near Cincinnati, Oh. M.I. Oct. 22, 1861.

Commander: Capt. Lewis Heckman. b. Germany around 1823. Residence Cleveland, occupation "manufacturer fancy cakes and confectionery." 2 Lt., Battery K, May 1, 1863. Not considered an effective commander. M.O. July 12, 1865. d. Lakeview House, Rocky River, Oh., Aug. 1, 1872.

Number: 4 12-pdr. Napoleons. 118 men.   **Loss:** 2 k., 11 wd., 2 m.

Monument: Carlisle and Lincoln Streets, Gettysburg. Map reference **I G-13**

"Arriving about noon July 1, 1863, this battery, Captain Lewis Heckman commanding, went into position here in reserve. When the 11th Corps began to retire, it engaged the enemy with great gallantry. After severe loss it was withdrawn."

1st OHIO ARTILLERY, BATTERY L
5th Corps Artillery Brigade

Raised: Scioto County

Organized: Camp Dennison, near Cincinnati, Oh. M.I. Oct. 1861-Jan. 1862.

Commander: Capt. Frank Charles Gibbs. b. Honesdale, Penn., Sept. 15, 1836. Surveyor in Portsmouth, Oh. Pvt., 1st Ohio Infantry (3 mos.). Lt., Battery L, Nov. 1, 1861. Wd. Oct. 19, 1864. M.O. July 4, 1865. d. Tacony, Philadelphia County, Penn., Aug. 2, 1888.

Number: 6 12 pdr. Napoleons. 121 men.   **Loss:** 2 wd.

Monument: Sykes Ave. Map reference **V E-11**

"Arriving on the field at 8 a.m. July 2, went into position under a brisk skirmish fire on the extreme right on Wolf Hill. Afterwards moved to north slope of Little Round Top, and there became hotly engaged with Longstreet's Corps then trying to turn the left. Held same position July 3."

Marker: Right section flank marker north of Wheatfield Road. Map reference **V C-12**

1st OHIO CAVALRY, COMPANIES A AND C
Company A, Headquarters, 3rd Div., Cavalry Corps
Company C, Headquarters, 2nd Div., Cavalry Corps

Raised: Fayette County

Organized: Camp Chase, near Columbus, Oh. Only two companies of the regiment served in the eastern theater. In the spring of 1864, the companies were sent west to join the regiment serving with the Army of the Cumberland.

Commanders: Co. A, Capt. Noah Jones. b. Fayette County, Oh., Feb. 10, 1840. Farmer near Bloomsburg, Oh. 2 Lt., Co. A, 1st Ohio Cav., Aug. 16, 1861. M.O. Nov. 16, 1864. d. Washington Court House, Oh., May 11, 1902. Co. C, Capt. Samuel N. Stanford. b. Ohio, 1837 or 1838. Lawyer in Springfield, Oh. Pvt., 3rd Ohio Inf. (3 mos.).2 Lt., Co. C, 1st Oh. Cav., Sept. 12, 1861. Cashiered to date April 25, 1864 for drunkenness on duty. d. New Moscow, Coshocton County, Oh., June 26, 1871.

Number: 85   **Loss:** None

Monument: Taneytown Road. Map reference **II J-17**

"Companies A and C, First Ohio Cavalry, July 1, 2, 3, 1863. During the Battle of Gettysburg these companies furnished bearers of dispatches to different parts of the field. In the course of the campaign they several times vigorously engaged the enemy."

6th OHIO CAVALRY

Detached with its brig. in Maryland during the Battle of Gettysburg.

56th Pennsylvania Infantry, North Reynolds Avenue.

11th PENNSYLVANIA INFANTRY
1st Corps, 2nd Div., 2nd Brig.

Raised: Counties of Westmoreland, Clinton, Lycoming, Allegheny, Carbon and Cumberland.

Organized: Camp Curtin, Harrisburg, Penn. M.I. Sept.-Nov. 1861.

Commanders: Col. Richard Coulter. b. Greensburg, Penn., Oct. 1, 1827. Grad. Jefferson College, Penn., 1845. Mexican War service. Lawyer in Greensburg. Lt. Col., 11th Penn. (3 mos.). Col., 11th Penn. (3 yrs.), Nov. 27, 1861. Wd. Dec. 13, 1862, July 3, 1863 and May 18, 1864. M.O. July 1, 1865. d. Greensburg, Oct. 14, 1908. When Coulter promoted to brigade command at Gettysburg, Capt. Benjamin Franklin Haines took regiment. b. Easton, Penn., Nov. 30, 1835. Clerk, Dunning, Penn. Pvt., 11th Penn. (3 mos.). 2 Lt., Co. B, 11th Penn. (3 yrs.), Sept. 10, 1861. Wd. Second Bull Run and July 3, 1863. M.O. Lt. Col., July 1, 1865. d. Phillipsburg, N.J., Sept. 28, 1903. When Haines wd., Capt. John B. Overmyer took command. b. Lycoming County, Penn., 1842. Carpenter. Enrolled in Jersey Shore, Penn. Pvt., 11th Penn., Sept. 27, 1861. M.O. Major, July 1, 1865. d. Chicago, Illinois Jan. 27, 1903.

Number: 292 **Loss:** 5 k., 52 wd., 60 m.

Monument: Doubleday Ave. Map reference I D-7

Figure of dog on monument represents "Sallie," the regimental mascot.

23rd PENNSYLVANIA INFANTRY
6th Corps, 3rd Div., 1st Brig.

Other Names: "Birney's Zouaves"

Raised: Philadelphia, Penn.

Organized: Camp near Falls of the Schuylkill, near Philadelphia, Penn. M.I. Aug. 2, 1861.

Commander: Lt. Col. John Francis Glenn. b. Philadelphia, Nov. 2, 1829. Mexican War service. Printer in Philadelphia. Member of Militia. Capt., 23rd Penn. (3 mos.). Capt., Co. A, 23rd Penn. (3 yrs.). Aug. 8, 1861. M.O. Col., Sept. 8, 1864. d. Philadelphia, Jan. 8, 1905.

Number: 538 **Loss:** 1 k., 13 wd.

Monument: Slocum Ave. Map reference III E-5

"The regiment was placed in reserve in rear of this position at 9:30 a.m. of the 3d, and subsequently five companies advanced into the breast-works. During the heavy cannonade it moved with the brigade to support the 'left center.'"

26th PENNSYLVANIA INFANTRY
3rd Corps, 2nd Div., 1st Brig.

Raised: Philadelphia, Penn.

Organized: Philadelphia, Penn. M.I. May 27, 1861.

Commander: Maj. Robert Lewis Bodine. b. Northampton, Penn., May 30, 1832.

Militia service. Bookkeeper in Philadelphia. Sergt., Co. D, 26th Penn., May 27, 1861. M.O. Lt. Col., June 18, 1864. d. Philadelphia, Jan. 12, 1874.

Number: 365   **Loss:** 30 k., 176 wd., 7 m.

Monument: Emmitsburg Road. Map reference **IV C-7**

"July 2nd went into action here."

27th PENNSYLVANIA INFANTRY
9 Companies
11th Corps, 2nd Div., 1st Brig.

Raised: Philadelphia, Penn.

Organized: Camp Einstein, Camden, N.J. M.I. May 30 and 31, 1861. During the Gettysburg campaign, company F in Washington, D.C.

Commander: Lt. Col. Lorenz Cantador. b. Prussia, June 10, 1810. Prussian military officer. Came to U.S. in 1851. Merchant in Philadelphia. Maj., 27th Penn., Sept. 7, 1861. M.O. Nov. 16, 1863. d. Dec. 2, 1883.

Number: 324   **Loss:** 6 k., 29 wd., 76 m.

Monuments: Coster Ave. and East Cemetery Hill.

Map references **II J-2, I H-14**

East Cemetery Hill—"July 1, 1863. The regiment moved with the Brigade on the afternoon to N.E. side of Gettysburg where it became actively engaged covering the retreat of the Corps. It then withdrew to this position where after dark of the 2nd it assisted in repulsing a desperate assault of the enemy. It subsequently moved into the cemetery where it remained until the close of the battle."

28th PENNSYLVANIA.INFANTRY
12th Corps, 2nd Div., 1st Brig.
Company B, Provost Guard, 2nd Div., 12th Corps.

Raised: Philadelphia. Counties of Allegheny, Luzerne, Westmoreland and Carbon.

Organized: Camp Coleman, Oxford Park, Philadelphia, Pa. M.I. June 28, 1861.

Commander: Capt. John Hornbuckle Flynn. b. Waterford, Ireland, Mar. 10, 1819. Came to U.S. and entered U.S. military service, 1844-1849. Merchant in Philadelphia before the Civil War. Lt. and Adjt., 28th Penn., July 28, 1861. Wd. Feb. 12, 1865. M.O. Col., Nov. 3, 1865. d. Little Rock, Ark., Dec. 25, 1875. Buried Little Rock National Cemetery.

Number: 370   **Loss:** 3 k., 23 wd., 2 m.

Monuments: Slocum Ave. and near Rock Creek.

Map references **III F-4, III I-6**

Slocum Ave—"Arrived at 5 p.m. July 1st and went into position on the ridge north of Little Round Top. At 6:30 a.m. July 2nd moved to Culp's Hill where the regiment advanced to Rock Creek to support the skirmish line. At dark retired and moved with the Brigade. Returned at about 3 a.m. July 3rd and at 8 a.m. relieved the troops in the breast-works; was relieved in turn and again advanced and occupied the works from 4 p.m. to 10 p.m."

Rock Creek—"The regiment took position here July 2d about 8 a.m. Deployed as skirmishers and was engaged with the enemy during the day. Remained until 7 p.m. when it was ordered to rejoin First Brigade."

29th PENNSYLVANIA INFANTRY
12th Corps, 2nd Div., 2nd Brig.

Other Names: "Jackson Regiment"

Raised: Philadelphia, Penn.

Organized: Camp between Haddington and Hestonville, Philadelphia, Penn. M.I. July 1861.

Commander: Col. William Rickards, Jr. b. Philadelphia, Nov. 18, 1824. Manufactured jewelry in Philadelphia. Capt., Co. I, 29th Penn., July 9, 1861. Wd. May 25, 1862 and June 15, 1864. M.O. Nov. 6, 1864. Became a dentist. d. Franklin, Penn., May 25, 1900.

Number: 485   **Loss:** 15 k., 43 wd., 8 m.

Monument: Slocum Ave. Map reference III F-7

"July 2. Position of the Regiment. At 7 p.m. the Brigade was withdrawn, and on returning during the night found the enemy in these works. The regiment took position in rear of this line, with its right as indicated by the tablet erected to the left and rear; and from there a charge of the enemy at daylight of the 3rd was repulsed. After a contest of over seven hours, in which the Regiment participated, it re-occupied and held the works until the close of the battle."

Marker: Slocum Ave. Map reference III E-6

1st PENNSYLVANIA RESERVES
5th Corps, 3rd Div., 1st Brig.

Other Names: "30th Pennsylvania Infantry."

Raised: Counties of Chester, Lancaster, Delaware, Cumberland and Adams (Co. K raised in and around Gettysburg, Penn.).

Organized: Camp Wayne near Westchester, Penn. M.I. July 26, 1861.

Commander: Col. William Cooper Talley. b. New Castle County, Delaware, Dec. 11, 1831. Start of war published a newspaper in Norristown, Penn. Sold paper to recruit Co. F, 1st Penn. Reserves. Capt., Co. F, July 26, 1861. Wd. June 30, 1862. M.O. June 13, 1864. Lived and died in Washington, D.C. d. Oct. 20, 1903. Buried Arlington National Cemetery.

Number: 444   **Loss:** 8 k., 38 wd.

Monument: Ayres Ave. Map reference V B-9

"July 2d in the evening charged from the hill in rear to this position and held it until the afternoon of July 3d when the Brigade advanced through the woods to the front and left driving the enemy and capturing many prisoners."

2nd PENNSYLVANIA RESERVES
9 Companies
5th Corps, 3rd Div., 1st Brig.

Other Names: "31st Pennsylvania Infantry"
Raised: Philadelphia, Penn. Also Lancaster County.
Organized: Camp Washington, near Easton, Penn. M.I. May 27, 1861. Company I disbanded in Aug. 1861.
Commander: Lt. Col. George Abisha Woodward. b. Wilkes-Barre, Penn., Feb. 14, 1835. Grad. Trinity College, Conn., 1855. City Attorney of Milwaukee, Wisc., 1858-1859. Resident of Philadelphia at start of Civil War. Capt., Co. A, 2nd Penn. Res., Aug. 1, 1861. Wd. June 30, 1862. M.O. Aug. 29, 1863. Subsequent service in Veteran Reserve Corps and 45th U.S. Inf. Retired Mar. 20, 1879. d. Washington, D.C., Dec. 22, 1916. Buried Arlington National Cemetery.
Number: 273   **Loss:** 3 k., 33 wd., 1 m.
Monument: Ayres Ave. Map reference **V B-9**
"July 2d in the evening charged from the hill in rear to this position and held it until the afternoon of July 3d when the Brigade advanced through the woods to the front and left driving the enemy and capturing many prisoners."

5th PENNSYLVANIA RESERVES
5th Corps, 3rd Div, 3rd Brig.

Other Names: "34th Pennsylvania Infantry"
Raised: Counties of Lycoming, Northumberland, Huntingdon, Clearfield, Union, Centre, Bradford and Lancaster.
Organized: Camp Curtin, Harrisburg, Penn. M.I. June 20, 1861.
Commander: Lt. Col. George Dare. b. Penn., 1836. Militia service. 1860 was a store keeper in Huntingdon, Penn. Maj., 5th Penn. Res., June 20, 1861. Wd. Dec. 13, 1862. Killed May 6, 1864 in the Battle of the Wilderness, Va.
Number: 334   **Loss:** 2 wd.
Monument: Big Round Top. Map reference **V I-9**
"Occupied this position on the evening of July 2d and held it to the close of the battle."

6th PENNSYLVANIA RESERVES
5th Corps, 3rd Div., 1st Brig.

Other Names: "35th Pennsylvania Infantry"
Raised: Counties of Bradford, Columbia, Snyder, Wayne, Franklin, Montour, Dauphin, Tioga, and Susquehanna.
Organized: Camp Curtin, Harrisburg, Penn. M.I. July 27, 1861.
Commander: Lt. Col. Wellington Harry Ent. b. Light Street, Penn., Aug. 16, 1834. Grad. Dickinson Seminary, Penn., 1858. Also Albany Law School, N.Y., 1860. Lawyer in Bloomsburg, Penn. Militia service. Capt., Co. A, 6th Penn. Res., July 27, 1861. Wd. May 30, 1864. M.O. Col., June 11, 1864. d. Bloomsburg, Nov. 5, 1871.

Number: 380 **Loss:** 2 k., 22 wd.

Monument: N.E. of Wheatfield. Map reference **V A-9**

"July 2d in the evening charged from the hill in rear to this position and held it until the afternoon of July 3d when the Brigade advanced through the woods to the front and left driving the enemy and capturing many prisoners."

9th PENNSYLVANIA RESERVES
5th Corps, 3rd Div., 3rd Brig.

Other Names: "38th Pennsylvania Infantry"

Raised: Counties of Allegheny, Crawford and Beaver.

Organized: Camp Wright, near Pittsburgh, Penn. M.I. July 28, 1861.

Commander: Lt. Col. James M'Kinney Snodgrass. b. Penn., 1806. Militia officer. In 1860 he was a farmer in Hope Church, Penn. Maj., 9th Penn. Res., July 16, 1861. M.O. Mar. 29, 1864. His superiors wrote of him, "he is when with his regiment a careless and inefficient officer." d. June 14, 1883 at the home for aged men in Wilkinsburg, Penn.

Number: 377 **Loss:** 5 wd.

Monument: Warren Ave. Map reference **V G-10**

"The Regiment arrived on the field July 2d about 5 p.m. with 377 officers and men and soon after moved to this position and held it until the close of the battle with a loss of five wounded."

10th PENNSYLVANIA RESERVES
5th Corps, 3rd Div., 3rd Brig.

Other Names: "39th Pennsylvania Infantry"

Raised: Counties of Beaver, Mercer, Crawford, Somerset, Venango, Washington, Clarion and Warren.

Organized: Camp Wright, near Pittsburgh, Penn. M.I. July 21, 1861.

Commander: Col. Adoniram Judson Warner. b. Wales, near Buffalo, N.Y., Jan. 13, 1834. Headed the Union School of Mercer, Penn. Capt., Co. G, 10th Penn. Res., July 21, 1861. Wd. Sept. 17, 1862. Service in Veteran Reserve Corps until muster out Nov. 17, 1865. Served two terms in U.S. Congress. d. Marietta, Oh., Aug. 12, 1910.

Number: 420 **Loss:** 2 k., 3 wd.

Monument: Sykes Ave. Map reference **V H-10**

"July 2d occupied this line of stone fence and remained from 5 p.m. until the close of the battle."

11th PENNSYLVANIA RESERVES
5th Corps, 3rd Div., 3rd Brig.

Other Names: "40th Pennsylvania Infantry"

Raised: Counties of Indiana, Butler, Westmoreland, Fayette, Armstrong, Jefferson and Cambria.

Organized: Camp Wright, near Pittsburgh, Penn. M.I. June 29, 1861.

Commander: Col. Samuel McCartney Jackson. b. near Apollo, Penn., Sept. 24, 1833. Militia service. Merchant in Apollo. Capt., Co. G, 11th Penn. Res., June 8, 1861. M.O. June 13, 1864. Elected to state legislature. d. Apollo, Penn., May 8, 1907.

Number: 392 **Loss:** 3 k., 38 wd.

Monument: Ayres Ave. Map reference **V B-9**

"July 2d in the evening charged from the hill in rear to this position and held it until the afternoon of July 3d when the Brigade advanced through the woods to the front and left driving the enemy and capturing many prisoners."

12th PENNSYLVANIA RESERVES
9 Companies
5th Corps, 3rd Div., 3rd Brig.

Other Names: "41st Pennsylvania Infantry"

Raised: Philadelphia. Counties of Wyoming, Bradford, Dauphin, Northampton, Westmoreland, York, Indiana, Huntingdon, and Franklin.

Organized: Camp Curtin, Harrisburg, Penn. M.I. Aug. 10, 1861. Co. K absorbed into other companies in July 1862.

Commander: Col. Martin Davis Hardin. b. Jacksonville, Ill., June 26, 1937. Grad. West Point, 1859. Service in 3rd U.S. Artillery. Lt. Col., 12th Penn. Res., July 8, 1862. Wd. Aug. 29, 1862 and Dec. 14, 1863. M.O. Jan. 15, 1866. Service in 43rd U.S. Infantry. Retd. with rank of Brig Gen., Dec. 15, 1870. Lawyer in Chicago, Ill. d. St. Augustine, Florida, Dec. 12, 1923.

Number: 320 **Loss:** 1 k., 1 wd.

Monument: Big Round Top. Map reference **V J-9**

"Occupied this position on the evening of July 2d and held it to the close of the battle."

13th PENNSYLVANIA RESERVES
5th Corps, 3rd Div., 1st Brig.

Other Names: "42nd Pennsylvania Infantry." "Bucktails." "1st Rifles."

Raised: Counties of Tioga, Perry, Cameron, Warren, Carbon, Elk, Chester, McKean and Clearfield.

Organized: Camp Curtin, Harrisburg, Penn. M.I. without date.

Commanders: Col. Charles Frederick Taylor. b. West Chester, Penn., Feb. 6, 1840. Attnd. University of Michigan. Farmer in Kennett Square, Penn. Capt., Co. H, 13th Penn. Res., July 22, 1861. Wd. Dec. 13, 1862. Killed July 2, 1863 at Gettysburg. Maj. William Ross Hartshorne then assumed command. b. Curwensville, Penn., Jan. 26, 1839. Lumberman in Curwensville, Penn. Lt., Co. K, 13th Penn. Res., June 21, 1861. Wd. June 26, 1862. M.O. June 11, 1864. Col., 190th Penn. Inf. M.O. June 28, 1865. d. Philadelphia, Penn., June 12, 1905.

Number: 349 **Loss:** 7 k., 39 wd., 2 m.

Monument: Ayres Ave. Map reference **V C-8**

"July 2d in the evening charged from the hill in rear to this position and held it until the afternoon of July 3d when the Brigade advanced through the woods to the front and left driving the enemy and capturing many prisoners."

Marker: Near monument. Map reference V C-8 Indicates where Col. Taylor fell. Current marker replaced original in 1905.

46th PENNSYLVANIA INFANTRY
12th Corps, 1st Div., 1st Brig.

Raised: Counties of Allegheny, Northampton, Potter, Mifflin, Berks, Luzerne and Dauphin.

Organized: Camp Curtin, Harrisburg, Penn. M.I. Oct. 31, 1861.

Commander: Col. James Levan Selfridge. b. Berks County, Penn., Sept. 22, 1824. Attend. Lafayette College, Penn. Coal and real estate business in Bethlehem, Penn. Capt., 1st Penn. Inf. (3 mos.). Lt. Col., 46th Penn., Aug. 8, 1861. Wd. Aug.9, 1862. M.O. July 16, 1865. "He was particularly adapted to military life but not to private occupations." Committed suicide in Philadelphia, Penn., May 19, 1887.

Number: 262 **Loss:** 2 k., 10 wd., 1 m.

Monument: Slocum Ave. Map reference III G-8

"July 2. The Regiment constructed and held these works until evening when the Division moved to support the left of the line. Returning in the night the enemy was found in the works and the Regiment was posted in the open field in the rear until the enemy was driven out, when it returned and held the works until the close of the battle.

July 3, 1863 p.m. ordered to support of the centre between General Meade's headquarters and the fighting line and in reserve. After repulse of Longstreet's assault returned to breastwork. July 4, a.m. Reconnoitered towards Hanover. Returned through Gettysburg and encamped."

49th PENNSYLVANIA INFANTRY
4 Companies, ABCD
6th Corps, 1st Div., 3rd Brig.

Raised: Counties of Huntingdon, Mifflin, Chester, Centre and Juniata.

Organized: Camp Curtin, Harrisburg, Penn. M.I. Oct. 24, 1861.

Commander: Lt. Col. Thomas Marcus Hulings. b. Lewistown, Penn., Feb. 7, 1835. Lawyer. Lt., 25th Penn. Inf. (3 mos.). Capt., 12th U.S. Infantry. Maj., 49th Penn., July 31, 1861. Killed leading regiment as its Col., May 10, 1864, in the Battle of Spotsylvania Court House, Va.

Number: 318 **Loss:** None

Monument: Howe Ave. Map reference V L-17

"This regiment made a continuous march from Manchester Md. arriving on the field the afternoon of July 2. Occupied this position in reserve from the morning of the 3rd until the enemy's assault in the afternoon when it moved to support centre thence to Round Top."

53rd PENNSYLVANIA INFANTRY
2nd Corps, 1st Div., 4th Brig.
Companies ABK,
2nd Corps Provost Guard

Raised: Counties of Montgomery, Luzerne, Potter, Northumberland, Juniata and Westmoreland.
Organized: Camp Curtin, Harrisburg, Penn. M.I. Nov. 7, 1861.
Commander: Lt. Col. Richards McMichael. b. Robeson Township, Berks County, Penn., Feb. 21, 1816. Mexican War service. Carpenter in Reading, Penn. Lt. Col., 14th Penn. Inf. Lt. Col., 53rd Penn., Nov. 7, 1861. M.O. May 17, 1864. Lt. Col., 194th Penn. M.O. Nov. 5, 1864. d. Dec. 5, 1894.
Number: 136   **Loss:** 7 k., 67 wd., 6 m.
Monument: Brooke Ave. Map reference V C-4
"July 2, about 5 p.m. the regiment deployed with the Brigade on the northerly side of, and charged through, the Wheatfield driving the enemy, and continuing the advance to this position, holding it until ordered to retire. July 3, in position with Division on left centre."

56th PENNSYLVANIA INFANTRY
9 Companies
1st Corps, 1st Div., 2nd Brig.

Raised: Philadelphia. Counties of Indiana, Luzerne, Centre and Susquehanna.
Organized: Camp Curtin, Harrisburg, Penn. M.I. Mar. 7, 1862. No company E at Gettysburg.
Commander: Col. John William Hofmann. b. Philadelphia, Feb. 18, 1824. Militia service. Merchant in Philadelphia. Capt., 23rd Penn. Infantry (3 mos.). Lt. Col., 56th Penn., Oct. 1, 1861. M.O. Col., Mar. 8, 1865. d. Philadelphia, Mar. 5, 1902.
Number: 252   **Loss:** 14 k., 61 wd., 55 m.
Monument: Reynolds Ave. Map reference I F-5
"The Regiment here delivered the opening fire of the infantry in the Battle of Gettysburg in the forenoon of July 1st 1863.
July 2 & 3. Occupied position on Culp's Hill as indicated by stone markers."
Markers: Culp's Hill. Markers not found.

57th PENNSYLVANIA INFANTRY
8 Companies
3rd Corps, 1st Div., 1st Brig.

Raised: Counties of Bradford, Mercer, Crawford and Tioga.
Organized: Camp Curtin, Harrisburg, Penn. M.I. Dec. 14, 1861. Companies D and G disbanded in Sept. 1862.
Commanders: Col. Peter Sides. b. Philadelphia, 1820. Merchant in Phila. Capt., Co. A, 57th Penn., Dec. 7, 1861. Wd. July 2, 1863 and May 5, 1864. M.O. Nov.

28, 1864. d. Phila., Oct. 23, 1878. When Sides wd., Capt. Alanson Henry Nelson took command. b. Tompkins County, New York, April 22, 1828. Farmer and lumberman in Titusville, Penn. Lt., Co. K, 57th Penn., Sept. 4, 1861. M.O. Nov. 4, 1864. d. Minneapolis, Minnesota, Jan. 14, 1921.

Number: 207 **Loss:** 11 k., 46 wd., 58 m.

Monument: Emmitsburg Road. Map reference **IV H-2**

"The Regiment occupied this position, exposed to a heavy artillery fire on the afternoon of July 2, for two hours, when it advanced 170 feet and engaged the enemy."

61st PENNSYLVANIA INFANTRY
6th Corps, 2nd Div., 3rd Brig.

Raised: Philadelphia. Counties of Allegheny, Luzerne, and Indiana.

Organized: Camp Copeland, Pittsburgh, Penn. M.I. Aug. 1861.

Commander: Lt. Col. George Fairlamb Smith. b. West Chester, Penn., Feb. 28, 1840. Grad. Yale College, 1858 and was studying law in West Chester when war began. Pvt., 2nd Penn. Inf. (3 mos.). Capt., 49th Penn. Inf. Maj., 61st Penn., Mar. 15, 1862. Wd. May 31, 1862 and May 12, 1864. M.O. Col., April 26, 1865. d. West Chester, Oct. 18, 1877.

Number: 400 **Loss:** 1 wd., 1 m.

Monument: Neill Ave. Map reference **III O-16**

"After a march of 37 miles reached the field about 4 p.m. July 2nd and moved to support of 12th Corps. Occupied this position from morning of July 3rd until close of battle."

62nd PENNSYLVANIA INFANTRY
12 Companies
5th Corps, 1st Div., 2nd Brig.

Other Names: "33rd Independent Pennsylvania Regiment"

Raised: Counties of Allegheny, Clarion, Armstrong, Blair and Jefferson.

Organized: Pittsburgh, Penn. M.I. July 4, 1861.

Commander: Lt. Col. James C. Hull. b. Pittsburgh, 1828. Mexican War service. Militia service. Carpenter in Allegheny City, Penn. Capt., Co. A, 62nd Penn., July 4, 1861. Mortally wounded May 12, 1864 at Battle of Spotsylvania Court House, Va. d. May 22, 1864.

Number: 426 **Loss:** 28 k., 107 wd., 40 m.

Monument: De Trobriand Ave. Map reference **V C-6**

"Position occupied by the Regiment on the evening of July 2, 1863 after the troops on the right had retired, and where the Brigade had a bayonet contest."

63rd PENNSYLVANIA INFANTRY
3rd Corps, 1st Div., 1st Brig.

Raised: Counties of Allegheny and Clarion.

Organized: Camp Haus, Washington, D.C. M.I. October 1861.

Commander: Maj. John Anderson Danks. b. Venango County, Penn., Mar. 11, 1826. Iron worker in Etna, Penn. Capt., Co. E, 63rd Penn., Sept. 9, 1861. Wd. May 31, 1862 and May 5, 1864. M.O. Aug. 5, 1864. Elected to state legislature. d. Glenfield, Penn., July 25, 1896.

Number: 296 **Loss:** 1 k., 29 wd., 4 m.

Monument: Emmitsburg Road. Map reference **IV K-1**

"The Regiment arrived on the battlefield about 8 p.m. July 1st and was immediately deployed upon picket 300 yards north of the Emmitsburg Road and in front of this position. Skirmish firing was kept up on the 2d from early morning until 5:30 p.m. when the regiment was relieved and rejoined the Brigade. On the 3rd in position on left center." The preceding inscription was written for the monument but does not appear there.

68th PENNSYLVANIA INFANTRY
3rd Corps, 1st Div., 1st Brig.

Other Names: "Scott Legion"

Raised: Philadelphia. Also Montgomery County.

Organized: Frankford, Philadelphia, Penn. M.I. Sept. 2, 1862.

Commanders: Col. Andrew Hart Tippin. b. Plymouth, Penn., Dec. 25, 1822. Mexican War service. Deputy marshal in Philadelphia. Maj., 20th Penn. Inf. (3 mos.). Col., 68th Penn., Sept. 1, 1862. M.O. June 9, 1865. d. Philadelphia, Feb. 6, 1870. When Tippin took command of brigade at Gettysburg, Capt. Milton S. Davis commanded regiment. b. Chester County, Penn., 1824. Mexican War Service. Carpenter in Philadelphia. Lt., 20th Penn. Inf. (3 mos.). Capt., Co. F, 68th Penn., Aug. 23, 1862. Killed Nov. 27, 1863 during the Mine Run Campaign.

Number: 394 **Loss:** 13 k., 126 wd., 13 m.

Monuments: Peach Orchard and Wheatfield Road. Map references **IV L-1, IV K-2**

Wheatfield Road—"This monument marks the left of the regiment while supporting Clark's Battery July 2d 1863. The right resting 150 feet north as indicated by flank marker. In the afternoon the Regiment advanced southward into the Peach Orchard where its other monument stands and engaged the enemy.

July 3d and 4th. The Regiment was in line with the Division on left centre."

69th PENNSYLVANIA INFANTRY
2nd Corps, 2nd Div., 2nd Brig.

Other Names: "Second California"

Raised: Philadelphia, Penn.

Organized: Camp Owen, Haddington, Philadelphia, Penn. M.I. Aug. 19, 1861. Formed from an Irish American militia company in Philadelphia. Accepted by Federal Government as part of Col. Edward Baker's "California Brigade."

When the state of Penn. claimed the units for its quota, they became known as the Philadelphia Brigade and were given state numbers.

Commanders: Col. Dennis O'Kane. b. Ireland, 1824. Tavern keeper in Philadelphia. Maj., 24th Penn. Infantry. Lt. Col., 69th Penn., Aug. 19, 1861. d. July 4, 1863 from wounds received on July 3, 1863. Capt. William Davis. b. Ireland, 1832 or 1833. Hatter in Philadelphia. Capt., Co. K, 69th Penn., Aug. 26, 1861. Wd. Aug. 25, 1864. M.O. Lt. Col., July 1, 1865. d. Philadelphia, Dec. 18, 1883.

Number: 329   **Loss:** 40 k., 80 wd., 17 m.

Monument: Webb Ave. Map reference **II E-12**

"This position was held by the 69th PA. Vols., July 2nd and 3rd 1863. Late on the afternoon of the 2nd, this regiment assisted in repulsing a desperate attack made by Wright's Ga. Brigade. About 1 O'Clock, p.m. of the 3rd, these lines were subjected to an artillery fire from nearly 150 guns, lasting over one hour after which, Pickett's Division charged this position, was repulsed, and nearly annihilated. The contest on the left and centre of this regiment, for a time being hand-to-hand. Of the regimental commanders attacking, but one remained unhurt. Genl. Garnett was killed, Genl. Kemper desperately wounded, and Genl. Armistead, after crossing the stonewall above the right of this command—2 companies of which changed front to oppose him—fell mortally wounded."

Markers: Markers on both sides of monument show the alignment of companies.

71st PENNSYLVANIA INFANTRY
2nd Corps, 2nd Div., 2nd Brig.

Other Names: "California Regiment"

Raised: Philadelphia and New York City.

Organized: Fort Schuyler, Throgs Neck, N.Y. City. M.I. June 1861. Formed in N.Y. City and Philadelphia, Penn. Accepted by the Federal Government as part of a possible future California quota. Later claimed by Pennsylvania as part of the Philadelphia brigade.

Commander: Col. Richard Penn Smith, Jr. Father famous playwright. Smith, Jr. b. Philadelphia, May 9, 1837. Attend. West Chester College, Penn. Clerk in Philadelphia. Lt., Co. F, 71st Penn., May 28, 1861. Wd. Sept. 17, 1862. M.O. July 2, 1864. d. Staten Island, N.Y., Nov. 27, 1887.

Number: 331   **Loss:** 21 k., 58 wd., 19 m.

Monument: Webb Ave. Map reference **II E-11**

"To the left of this point on July 2, the 71st Penna. assisted in repulsing the furious attack of Wright's Ga. Brig. During the terrific cannonading of July 3, the regiment occupied a position 60 yards in the rear of this spot, a number of the men voluntarily helping to work Cushing's disabled Battery. As the enemy emerged from Seminary Ridge the regiment was ordered forward, the left wing to this point, the right to the right in the rear. When Pickett's Division rushed upon the wing in overwhelming numbers it fell back into line with the right, thus bringing the whole regiment into action, with the additional use of a large number of loaded muskets gathered from the battle field of the previous day."

72nd PENNSYLVANIA INFANTRY
2nd Corps, 2nd Div., 2nd Brig.

Other Names: "Baxter's Philadelphia Fire Zouaves." "Third California."
Raised: Philadelphia, Penn.
Organized: Camp Lyon at Haddington, Philadelphia, Penn. M.I. Aug. 10, 1861. Recruited from volunteer fire companies in Philadelphia. Originally part of Col. Baker's "California Brigade." Later claimed by Pennsylvania as part of the "Philadelphia Brigade."
Commanders: Col. Dewitt Clinton Baxter. b. Dorchester, Mass., Mar. 9, 1829. Militia officer. Wood engraver in Philadelphia. Lt. Col., 19th Penn. Inf. (3 mos.). Col., 72nd Penn., Aug. 10, 1861. Wd. July 2, 1863 and May 6, 1864. M.O. Aug. 24, 1864. d. Philadelphia, May 9, 1881. When Baxter wd. at Gettysburg, Lt. Col. Theodore Hesser took command of regiment. b. Philadelphia, July 16, 1829. Mexican War service. Clerk in Philadelphia. Capt., 18th Penn. Infantry (3 mos.). Lt. Col., 72nd Penn., Aug. 10, 1861. Killed Nov. 27, 1863 during the Mine Run campaign.
Number: 458   **Loss:** 44 k., 146 wd., 2 m.
Monuments: Both along Webb Ave. Map references II E-12
Forward position monument_"July 2d 1863. The Regiment reached this angle at 1 a.m. Took position in rear of this monument. Supported Cushing's Battery A, 4th U.S. Artillery. At 6 p.m. assisted in repulsing an attack of the enemy and in making a counter-charge, driving them beyond the Emmitsburg Road capturing 250 prisoners.
July 3d 1863. The Regiment assisted in repulsing the charge of the enemy on the angle at 3 p.m. and in capturing many standards and prisoners. During the cannonading which preceded the charge the Regiment was in line sixty yards to the left and rear of this monument. When the rebels forced the troops from the first line the 72nd Regiment fought its way to the front and occupied the wall."

73rd PENNSYLVANIA INFANTRY
11th Corps, 2nd Div., 1st Brig.

Other Names: "Pennsylvania Legion." "45th Pennsylvania Volunteers."
Raised: Philadelphia, Penn.
Organized: Camp on Engle's and Wolf's farms at Lemon Hill, near Philadelphia, Penn. M.I. Sept. 19, 1861. Mar. 1863 many men from 66th Penn. Infantry added to the regiment.
Commander: Capt. Daniel F. Kelly. b. Ireland, 1837. Watchmaker in Philadelphia. Capt., Co. F, 73rd Penn., Sept. 11, 1861. M.O. June 22, 1864.
Number: 332   **Loss:** 7 k., 27 wd.

Monument: East Cemetery Hill. Map reference II K-3

"July 1st. The Regiment arrived on Cemetery Hill at 2 p.m. and at a later hour moved into the town near the square to cover the retreat of the Corps. July 2nd. In the morning took position in the Cemetery. At dusk moved hastily to this position and in a severe contest assisted in repulsing a desperate assault on these batteries.

July 3rd. Returned to its former position in the Cemetery and assisted in repulsing the enemy's final assault."

74th PENNSYLVANIA INFANTRY
9 Companies
11th Corps, 3rd Div., 1st Brig.

Other Names: "Thirty Fifth Regiment"

Raised: Allegheny County, Penn. Also Philadelphia, Penn. Company C not at Gettysburg.

Organized: Camp Wilkins, near Pittsburgh, Penn. M.I. Sept. 14, 1861.

Commanders: Col. Adolph Von Hartung. b. Kuestrin, Prussia, June 24, 1834. Officer in Prussian army. Came to U.S. and apparently settled in Baltimore, Md. Merchant at start of the war. Went to Philadelphia to raise a company of soldiers. Capt., Co. A, 74th Penn., Aug. 5, 1861. Wd. in leg July 1, 1863. M.O. July 11, 1864. For a time he ran a military school in Hamburg, Germany. Returned to Baltimore where he died April 10, 1902. When Von Hartung was wounded on July 1, he sent for Lt. Col. Alexander Theobald Von Mitzel to take command of regiment. b. Berlin, 1835. Prussian Army officer. Came to U.S. and settled in Baltimore, Md. Capt., Co. K, 74th Penn., Sept. 21, 1861. M.O. Oct. 15, 1864. Saloon business in Baltimore where he died, Sept. 26, 1887. When Von Mitzel was captured in the retreat of July 1, Capt. Henry Krauseneck led the regiment. Not much is known about his life. There does exist a U.S. passport application dated Nov. 19, 1864 made out by a Henry Krauseneck. It has b. Bavaria, Mar. 24, 1828. Soldier Krauseneck joined Co. D, 74th Penn. as 2 Lt. on Jan. 7, 1862. In Jan. 1864 he was brought up on charges of cowardice while in command of the regiment at Gettysburg. He was found guilty and allowed to resign May 24, 1864.

Number: 381 **Loss:** 10 k., 40 wd., 60 m.

Monument: Howard Ave. Map reference I E-12

"July 1st fought here from 2 p.m. until the Corps fell back. July 2d and 3d in line with Division in front of Cemetery."

Marker: National Cemetery. Map reference II H-4

75th PENNSYLVANIA INFANTRY
9 Companies
11th Corps, 3rd Div., 2nd Brig.

Other Names: "40th Pennsylvania Volunteers"

Raised: Philadelphia, Penn.

Organized: Camp Worth, Hestonville, Philadelphia, Penn. M.I. Aug. 1861. Co. K not organized until 1865.

Commanders: Col. Francis Mahler. b. Baden, Germany, 1826. Joined the 1848 revolution against the government. Fled to U.S. Cordwainer in Philadelphia. Lt. Col., 75th Penn., Aug. 27, 1861. Mort. wd., July 1, 1863. d. July 4. When Mahler wd., Maj. August Ledig took command. b. Merseburg, Prussia, 1816. Participated in the revolution of 1848. Came to U.S. in 1852. Machinist in Philadelphia. Capt., 21st Penn. Infantry (3 mos.). Capt., Co. E, 75th Penn., Sept. 6, 1861. M.O. July 30, 1865. d. Philadelphia, Aug. 12, 1895.

Number: 258 **Loss:** 19 k., 89 wd., 3 m.

Monuments: Howard Ave. and National Cemetery.

Map references **I D-13, II J-3**

Howard Ave—"July 1. Fought on this position from 2 p.m. until the Corps retired. July 2 & 3. Held position at stone wall near the Cemetery as shown by monument there."

81st PENNSYLVANA INFANTRY
2nd Corps, 1st Div., 1st Brig.

Raised: Philadelphia, Penn. Counties of Carbon and Luzerne.

Organized: Camp Washington, Easton, Penn. M.I. Aug.-Oct. 1861.

Commander: Lt. Col. Amos Stroh. Father Abraham came to Mauch Chunk, Penn. in 1821 when Amos was a few months old. Amos was a moulder in Mauch Chunk at the start of the war. Capt., Co. G, 81st Penn., Sept. 16, 1861. Resigned July 22, 1863. Resignation accepted with a statement that he was unfit for service. d. Mauch Chunk, Feb. 7, 1899.

Number: 190 **Loss:** 5 k., 49 wd., 8 m.

Monument: Wheatfield. Map reference **V B-8**

"Fought on this line in the afternoon of July 2d."

82nd PENNSYLVANIA INFANTRY
6th Corps, 3rd Div., 1st Brig.

Other Names: "31st Pennsylvania Volunteers"

Raised: Philadelphia. Also Allegheny County.

Organized: Washington, D.C. M.I. Sept. 8, 1861.

Commander: Col. Issac Clark Mifflin Bassett. b. Philadelphia, 1827. Mexican War service. Coal merchant in Philadelphia. Capt., 17th Penn. Inf. (3 mos.). Capt., Co. K, 82nd Penn., Sept. 21, 1861. Wd. June 3, 1864. M.O. July 20, 1865. d. Philadelphia, Oct. 2, 1869.

Number: 320 **Loss:** 6 wd.

Monument: Slocum Ave. Map reference **III F-5**

"July 3rd marched from near Little Round Top and occupied the works in front at 11:30 a.m. relieving other troops."

83rd PENNSYLVANIA INFANTRY
5th Corps, 1st Div., 3rd Brig.

Raised: Counties of Erie, Crawford and Forest.
Organized: Camp McLane near Erie, Penn. M.I. Sept. 8, 1861.
Commander: Capt. Orpheus Saeger Woodward. b. Harbor Creek, Penn., May 1, 1835. School teacher there at time of enlistment. Capt., Co. D, 83rd Penn., Sept. 13, 1861. Wd. May 5, 1864. Right leg amputated. M.O. Col., Sept. 20, 1864. Served two terms in Penn. legislature. d. Leavenworth, Kansas, June 26, 1919.
Number: 308   **Loss:** 10 k., 45 wd.
Monument: Sykes Ave. Map reference **V G-11**
"The Brigade was hurried to Little Round Top about 5 p.m. of July 2d. This Regiment taking position in front of this monument and repulsed several desperate charges of the enemy after which this Regiment assisted in driving the enemy beyond and in taking possession of Big Round Top. On the morning of the 3rd, rejoined the Brigade on the left centre."

84th PENNSYLVANIA INFANTRY

Not at Gettysburg. Guarding Trains at Westminster, Maryland.

88th PENNSYLVANIA INFANTRY
1st Corps, 2nd Div., 2nd Brig.

Other Names: "Cameron Light Guards"
Raised: Philadelphia. Also Berks County.
Organized: Camp Stokley, within modern Fairmount Park, Philadelphia, Penn. M.I. Sept. 1861.
Commanders: Maj. Benezet Forst Foust. b. Philadelphia, April 5, 1840. Lawyer in Philadelphia. Lt. and Adjt., Co. A, 88th Penn., Oct. 3, 1861. Wd. July 1, 1863. Resigned Nov. 7, 1863 to accept commission in Invalid Corps. M.O. Lt. Col., June 30, 1865. d. Philadelphia, Jan. 8, 1870. Capt. Edmund A. Mass took command from the wounded Foust. b. Reading, Penn., 1834. Baggage master on the Reading R.R. Lt., Co. B, 88th Penn., Sept. 13, 1861. Captured July 1, 1863. M.O. Lt. Col., June 30, 1865. d. Collingdale, Penn., Sept. 4, 1894. Finally, Capt. Henry Whiteside led the regiment. Place of birth in dispute, probably Reading, Penn., in 1835. Cooper in Reading. Pvt. 25th Penn. Inf., (3 mos.). Lt., Co. A, 88th Penn., Nov. 25, 1862. Wd. Sept. 17, 1862. M.O. Sept. 29,1864. Killed by train near Mays Landing, N.J., April 17, 1905.
Number: 296   **Loss:** 4 k., 55 wd., 51 m.
Monument: Doubleday Ave. Map reference **I C-7**
"About noon, July 1st 1863, the regiment was in line along the Mummasburg Road, 200 yards S.E. of this monument. Later it changed direction and formed here, charged forward and captured two battle flags and a number of prisoners. At 4 p.m. Division was overpowered and forced through the town.

July 2d the regiment was in position facing the Emmittsburg [sic] Road and on July 3rd at Ziegler's Grove, as indicated by markers."

Markers: 250 feet in front of monument, near Cyclorama Center, and along Hancock Ave. Map references **I C-7, II G-8, IV G-13**

90th PENNSYLVANIA INFANTRY
1st Corps, 2nd Div., 2nd Brig.

Raised: Philadelphia, Penn.

Organized: Oxford Park, Philadelphia, Penn. M.I. Oct. 1, 1861.

Commanders: Col. Peter Lyle. b. Philadelphia, Dec. 24, 1821. Militia service. Tobacconist in Philadelphia. Col., 19th Penn. Inf. (3 mos.). Col., 90th Penn., Mar. 10, 1862. M.O. Nov. 26, 1864. d. Philadelphia, July 17, 1879. When Col. Lyle took over brigade, Maj. Alfred Jacob Sellers commanded regiment. b. Plumsteadville, Penn., Mar. 2, 1836. Militia service. Broker in Philadelphia. Capt., 19th Penn. Inf. (3 mos.). Maj., 90th Penn., Feb. 26, 1862. Wd. Dec. 13, 1862. Awarded Medal of Honor for gallantry while leading his regiment at Gettysburg, July 1, 1863. M.O. Feb. 29, 1864. d. Philadelphia, Sept. 20, 1908.

Number: 208 **Loss:** 8 k., 45 wd., 40 m.

Monuments: Doubleday Ave. and near Cyclorama Center. Map references **I C-7, II F-8**
Cyclorama Center—"This monument marks the position of the 90th Penna. Volunteers of Philadelphia, July 3rd, 1863, Col. Peter Lyle, commanding the 1st Brigade, Major A. J. Sellers, the regiment. July 1st from one to three O'Clock p.m., the regiment fought on the extreme right of the 1st Corps on Seminary (Oak) Ridge as indicated by its monument there. Eight companies being refused, facing the Mummasburg Road. It there engaged Page's Va. Confederate Battery and O'Neal's Ala. Brigade of Rode's Division until its ammunition was exhausted. July 2nd it occupied Cemetery Hill and in the evening moved to left of 2nd Corps. Returning during the evening to this position."

Markers: Christ Lutheran Church in Gettysburg where Chaplain Horatio Stockton Howell was killed and along Hancock Ave. Map references **I J-12, IV G-13**

91st PENNSYLVANIA INFANTRY
5th Corps, 2nd Div., 3rd Brig.

Raised: Philadelphia, Penn.

Organized: Camp Chase east of Darby Road at 51st Street, Philadelphia, Penn. M.I. Dec. 4, 1861.

Commander: Lt. Col. Joseph Hill Sinex. b. near Stanton, Delaware, Oct. 7, 1819. Militia service. Carpenter in Philadelphia. Capt., 17th Penn. Inf. (3 mos.). Capt., Co. D, 91st Penn., Oct. 7, 1861. Wd. May 12, 1864. M.O. July 10, 1864. Elected to state legislature. d. Philadelphia, Oct. 6, 1892.

Number: 258 **Loss:** 3 k., 16 wd.

Monument: Sykes Ave. Map reference **V F-11**

"July 2. Moving at double-quick in the evening, the Regiment took position here and having aided in repulsing the attack of the enemy upon this line, remained until the close of the battle."

93rd PENNSYLVANIA INFANTRY
6th Corps, 3rd Div., 3rd Brig.

Raised: Counties of Lebanon, Berks and Montour.
Organized: Camp Coleman, Lebanon, Penn. M.I. Oct. 28, 1861.
Commander: Maj. John Irwin Nevin. b. Allegheny City, Penn., Aug. 23, 1837. Grad. Jefferson College, Penn., 1858. Teacher in Sewickley, Penn. 2 Lt., 28th Penn. Inf. Capt., Battery H, Penn. Light Artillery. Maj., 93rd Penn., April 1, 1863. M.O. Oct. 27, 1864. d. Sewickley, Jan. 5, 1884.
Number: 270 **Loss:** 10 wd.
Monuments: Sedgwick Ave. and north of Wheatfield Road. Map references **V B-11, V C-12**
Sedgwick Ave—"After charging with the Brigade from the right of Little Round Top in the evening of July 2d and assisting in the repulse of the enemy and in the capture of a number of prisoners, the Regiment retired to and held this position until after the close of the battle."
Near Wheatfield Road—"93rd Regiment Penn. Volunteers formed line of battle at this point under the immediate direction of Maj. Gen. John Sedgwick, commander of the 6th Corps evening of July 2, 1863 and advanced against the enemy taking the position indicated by monument at the foot of this hill where it remained until the close of the battle."

95th PENNSYLVANIA INFANTRY
6th Corps, 1st Div., 2nd Brig.

Other Names: "Gosline's Pennsylvania Zouaves." "45th Pennsylvania Volunteers." "54th Pennsylvania Volunteers."
Raised: Philadelphia, Penn.
Organized: Camp Gibson, Jones Woods, near Hestonville, Philadelphia, Penn. M.I. Sept.-Oct., 1861.
Commander: Lt. Col. Edward Carroll. b. Penn., 1825 or 1826. Carpenter in Philadelphia. Capt., Co. F, 95th Penn., Sept. 27, 1861. Wd. June 27, 1862. Killed May 5, 1864 during the Battle of the Wilderness, Va.
Number: 356 **Loss:** 1 k., 1 wd.
Monument: Wheatfield Road. Map reference **V C-11**
"Occupied this position in reserve from evening of July 2nd to morning of July 5th."

96th PENNSYLVANIA INFANTRY
6th Corps, 1st Div., 2nd Brig.

Raised: Schuylkill County

Organized: Camp on Lawton's Hill, Pottsville, Penn. Formed from National Light Infantry of Pottsville. M.I. Sept. 23, 1861.

Commander: Maj. William H. Lessig. b. Lebanon, Penn., Oct. 30, 1831. Militia service. Engineer and hotel keeper in Pottsville. Capt., Co. C, 96th Penn., Sept. 23, 1861. M.O. Lt. Col., Oct. 21, 1864. d. at Soldiers and Sailors Home at Monte Vista, Colorado, July 18, 1910.

Number: 356 **Loss:** 1 wd.

Monument: Wheatfield Road. Map reference **V C-11**

"Position of the 96th Regt. Penna. Volunteers, 2d Brigade, 1st Division, 6th Corps, from 5 p.m. of the 2d until the morning of the 5th of July 1863."

98th PENNSYLVANIA INFANTRY
6th Corps, 3rd Div., 3rd Brig.

Raised: Philadelphia, Penn.

Organized: Camp Ballier, near Girard College, Philadelphia, Penn. M.I. Aug.-Sept., 1861.

Commander: Maj. John Benedict Kohler. b. Germany, 1819. Stove manufacturer in Philadelphia. Capt., 21st Penn. Inf., (3 mos.). Capt., Co. K, 98th Penn., Sept. 12, 1861. Wd. June 11, 1864. Killed, Lt. Col., at the Battle of Cedar Creek, Va., Oct. 19, 1864.

Number: 406 **Loss:** 11 wd.

Monuments: Field north of Wheatfield Road and Sykes Ave. Map references **V B-11, V E-12**

Sykes Ave—"Arrived here July 2d about 5 p.m. Immediately charged to the Wheatfield and woods to the left. About dark rejoined the Brigade north of the road where other monument stands."

Wheatfield Road—"The Regiment was the advance of the Sixth Corps in its march from Manchester, Md. to the battlefield and occupied this position from the evening of July 2d until the close of the battle."

99th PENNSYLVANIA INFANTRY
3rd Corps, 1st Div., 2nd Brig.

Other Names: "32nd Pennsylvania Volunteers"

Raised: Philadelphia. Also Lancaster County.

Organized: Camp Franklin, near Alexandria, Virginia. M.I. July 1861-Jan. 1862.

Commander: Maj. John William Moore. b. Philadelphia, 1836. Lived in Philadelphia at the start of the war. Capt., Co. G, 66th Penn., Aug. 9, 1861. Company transferred to 99th Penn. and became Company K. Col., 203nd Penn., Sept. 10, 1864. Killed at Fort Fisher, N.C., Jan. 15, 1865.

Number: 339 **Loss:** 18 k., 81 wd., 11 m.

Monuments: Sickles Ave. and Hancock Ave. Map references **V E-7, II F-12**

Sickles Ave—"Fought on this line in the afternoon of July 2nd."

102nd PENNSYLVANIA INFANTRY
Detachment 6th Corps, 3rd Div., 3rd Brig.

Raised: Allegheny County
Organized: Pittsburgh, Penn. M.I. Aug. 16, 1861.
Commander: Lt. Robert Wilson Lyon. b. Butler County, Penn., 1842 or 1843. Blacksmith in Butler County at the time of the war. Pvt., 13th Penn. Inf. (3 mos.). Sergt., Co. H, 102nd Penn., Aug. 20, 1861. Wd. May 31, 1862 and Sept. 19, 1864. M.O. Capt., June 28, 1865. d. 1904.
Number: 103 **Loss:** None
Monument: North of Wheatfield Road. Map reference **V B-11**
"July 1. The Regiment was detailed at Manchester to guard trains to Westminster. At the latter place a detachment of 3 officers and 100 men was sent to Gettysburg with the supply train and on its arrival the morning of the 3d was posted on this line. The rest of the Regiment picketed the roads leading from Westminster to Gettysburg until the close of the battle."

105th PENNSYLVANIA INFANTRY
3rd Corps, 1st Div., 1st Brig.

Other Names: "Wild Cat Regiment"
Raised: Counties of Jefferson, Allegheny, Westmoreland, and Indiana.
Organized: Camp Wilkins, near Pittsburgh, Penn. M.I. Sept. 9, 1861.
Commander: Col. Calvin Augustus Craig. b. Greenville, Penn., Dec. 7, 1833. Merchant in Greenville. Capt., Co. C, 105th Penn., Aug. 28, 1861. Wd. July 2, 1863 and May 5, 1864. d. Aug. 17, 1864 of wounds received Aug. 16, 1864 at Deep Bottom, Va. Buried on a hill overlooking his boyhood home.
Number: 307 **Loss:** 8 k., 115 wd., 9 m.
Monument: Emmitsburg Road. Map reference **IV H-2**
"July 2nd. Position from 2 to 4 p.m. Moved across the Emmitsburg road. Being outflanked the Regiment changed front facing south and formed line along the lane at right angles to the road from which it retired fighting."

106th PENNSYLVANIA INFANTRY
2nd Corps, 2nd Div., 2nd Brig.

Other Names: "Fifth California"
Raised: Philadelphia. Counties of Bradford and Lycoming. Organized Camp at Bull's Head, West Philadelphia, Penn. M.I. Aug.-Sept., 1861. Originally part of Col. Baker's "California Brigade." Later claimed by Pennsylvania as part of the "Philadelphia Brigade."
Commander: Lt. Col. William Lovering Curry. b. Philadelphia, Jan. 29, 1833. Militia service. Paperhanger in Philadelphia, 2 Lt., 22nd Penn. Inf. (3 mos.). Lt. Col., 106th Penn., Aug. 28, 1861. Wd. May 11, 1864 at Spotsylvania Court House, Va. d. of wounds July 7, 1864.
Number: 335 **Loss:** 9 k., 54 wd., 1 m.

Monuments: Hancock Ave. and Emmitsburg Road. Map references **II E-12, II B-13**

Hancock Ave—"Position of the Regiment July 2, 1863. In the evening the Regiment assisted in repulsing a charge of the enemy on this line and made a counter charge to the Emmitsburg road in which 3 guns of Battery B, 1st Rhode Island were recovered and at the Codori House captured 250 prisoners.

The evening of July 2nd the Regiment moved to East Cemetery Hill to reinforce the 11th Corps and remained there as indicated by monument. During the 3rd, companies A and B continued here and assisted in repulsing the final assault of the enemy on the afternoon of the 3rd."

Marker: East Cemetery Hill. Map reference **II K-2**

107th PENNSYLVANIA INFANTRY
1st Corps, 2nd Div., 1st Brig.

Raised: Counties of Franklin, York, Dauphin, Cumberland and Lebanon.
Organized: Harrisburg, Penn. M.I. Feb. and Mar., 1862.
Commanders: Lt. Col. James McLean Thomson. b. Adams County, Penn., Feb. 4, 1834. Teacher and student of law. Capt., Co. B, 107th Penn., Feb. 15, 1862. Wd. July 1, 1863. M.O. July 13, 1865. Service in Veteran Volunteers until 1866. d. St. Louis, Missouri, Feb. 20, 1893. When Thomson wd., Capt. Emanuel D. Roath took command. b. Lancaster, Penn., Oct. 4, 1820. Member of state legislature. Start of war was magistrate of Marietta, Penn. Capt., Co. E, 107th Penn., Feb. 19, 1862. M.O. Mar. 5, 1865. d. Sept. 12, 1907.
Number: 255 **Loss:** 11 k., 56 wd., 98 m.
Monument: Doubleday Ave. Map reference **I D-7**

"July 1. The Regiment fought here from 1 p.m. until the Corps retired and then took position on the left of Cemetery Hill. In the evening of the 2nd moved to the left to support the Second Corps, and after the repulse of the enemy returned to former position. On the 3rd moved several times to reinforce different parts of the line."

Marker: Hancock Ave. Map reference **II F-9**

109th PENNSYLVANIA INFANTRY
12th Corps., 2nd Div., 2nd Brig.

Other Names: "Curtin Light Guards"
Raised: Philadelphia, Penn.
Organized: Camp at Nicetown, Philadelphia, Penn. M.I. Mar.-May, 1862.
Commander: Capt. Frederick Louis Gimber. b. New York, Mar. 12, 1836. Clerk in Philadelphia. Sergt., 19th Penn. Inf., (3 mos.). Capt., Co. E, 109th Penn., May 6, 1862. Wd. June 15, 1864. Rose to rank of Lt. Col. M.O. July 19, 1865 from the 111th Penn. Inf. d. Philadelphia, Oct. 14, 1910.
Number: 149 **Loss:** 3 k., 6 wd., 1 m.
Monument: Slocum Ave. Map reference **III F-6**

"July 1st. The Regiment arrived within two miles of Gettysburg about 5 p.m. and took position on the left of the Baltimore Pike. July 2nd it moved here and built these works. In the evening it was withdrawn with the Brigade, and returning in the night, found the works in the possession of the enemy, when it formed at right angles to this line behind a ledge of rocks to the left and rear of this position designated by a marker. After severe fighting on the morning of the 3rd this line was re-captured and held until the close of the battle."
Marker: Not found.

110th PENNSYLVANIA INFANTRY
6 Companies, ABCEHI
3rd Corps, 1st Div., 3rd Brig.

Raised: Philadelphia. Counties of Blair and Huntingdon.
Organized: Camp Curtin, Harrisburg, Penn. M.I. Oct. 24, 1861.
Commanders: Lt. Col. David Mattern Jones. b. Huntingdon County, Penn., April 24, 1838. Potter in Tyrone, Penn. Corpl., 3rd Penn. Inf. (3 mos.). Capt., Co. A, 110th Penn., Dec. 5, 1861. Wd. 2nd Bull Run campaign. Wd. July 2, 1863 resulting in amputation of his left leg. M.O. Oct. 9, 1863. d. Denver, Colorado, July 16, 1877. When Jones wd., Maj. Isaac Rodgers took command. b. Pennsylvania, Nov. 5, 1834. At start of war lived on family farm in Cromwell Township, Huntingdon county, Penn. Lt., Co. B, 110th Penn., Oct. 24, 1861. Mort. wd. as Lt. Col., May 12, 1864 at Spotsylvania Court House, Va. d. May 23, 1864. Last words to his family-"Tell them I have fought and fallen for my country."
Number: 152 **Loss:** 8 k., 45 wd.
Monument: De Trobriand Ave. Map reference **V B-5**
"July 2d. The Regiment fought on this line from 4 until 8 O'Clock p.m. July 3d supported batteries on Cemetery Hill."

111th PENNSYLVANIA INFANTRY
12th Corps, 2nd Div., 2nd Brig.

Raised: Counties of Erie, Warren and Crawford.
Organized: Camp Reed on the fair grounds near Erie, Penn. M.I. Jan. 24, 1862.
Commander: Lt. Col. Thomas McCormick Walker. b. Butler County, Penn., Feb. 4, 1834. Attnd. Princeton College, N.J. Civil Engineer in Erie, Penn. Maj. 111th Penn., Dec. 15, 1861. Wd. Oct. 28 or 29, 1863. M.O. Col., July 19, 1865. d. Fargo, North Dakota, April 6, 1910.
Number: 259 **Loss:** 5 k., 17 wd.
Monument: Slocum Ave. Map reference **III F-6**
"The Regiment built these works. In the evening of July 2 it was withdrawn with the Brigade, and returning during the night found the enemy in the works. Assisted in repulsing a charge of the enemy at daylight of the 3rd and after seven hours and a half of continuous fighting in which it participated, regained the works and held them until the close of the battle."

114th PENNSYLVANIA INFANTRY
3rd Corps, 1st Div., 1st Brig.

Other Names: "Collis Zouaves"
Raised: Philadelphia, Penn.
Organized: Camp Banks, Germantown, Philadelphia, Penn. M.I. Sept. 1, 1862.
Commanders: Lt. Col. Frederick Fernandez Cavada. b. Cuba, 1832. Living in Philadelphia at the start of the war. When asked occupation he said, "Nothing. I was engaged on the survey of the Panama R.R. but since I returned home have been unable to work because of my health." Capt., 23rd Penn. Inf. Lt. Col., 114th Penn., Sept. 1, 1862. Wd. Dec. 13, 1863. Captured at Gettysburg July 2, 1863. M.O. June 19, 1864. Appointed U.S. Consul to Cuba. Became Chief of General Staff of Cuban forces against Spain. Captured by Spanish forces and executed on July 1, 1871. Capt. Edward Roscoe Bowen took command after Cavada captured at Gettysburg. b. Philadelphia, Oct. 16, 1839. Clerk in Philadelphia. Pvt., Montgomery's Company of Heavy Artillery (3 mos.). 2 Lt., 75th Penn. Capt., Co. B, 114th Penn., Aug. 27, 1862. M.O. Lt. Col., May 29, 1865. d. Haverford, Penn., April 6, 1908.
Number: 312 **Loss:** 9 k., 86 wd., 60 m.
Monuments: Emmitsburg Road (July 2nd) and Hancock Ave. (July 3rd, 3 p.m.). Map references **IV 1-2, II G-11**

115th PENNSYLVANIA INFANTRY
9 Companies
3rd Corps, 2nd Div., 3rd Brig.

Raised: Philadelphia. Also Cambria County.
Organized: Hestonville, Philadelphia, Penn. M.I. Jan. 28, 1862. Company H never formed.
Commander: Maj. John Peter Dunne. b. Ireland, 1828. Carpenter in Philadelphia. Pvt. 24th Penn. Inf., (3 mos.). Capt., Co. B, 115th Penn., Jan. 28, 1862. M.O. Lt. Col., June 23, 1864. d. state hospital for insane at Norristown, Penn., Dec. 17, 1891.
Number: 182 **Loss:** 3 k., 18 wd., 3 m.
Monument: De Trobriand Ave. Map reference **V C-6**
"July 2. This Regiment detached from the Brigade engaged the enemy here at 4:30 p.m. July 3. In position with Division on left centre of the line."

116th PENNSYLVANIA INFANTRY
4 Companies, ABCD
2nd Corps, 1st Div., 2 Brig.
Company B, Provost Guard, 1st Div., 2nd Corps

Raised: Philadelphia, Penn.
Organized: Camp Emmet, Hestonville, Philadelphia, Penn. M.I. Sept. 1, 1862.
Commander: Maj. St. Clair Agustin Mulholland. b. Lisburn, County Antrim, Ireland, April 1, 1839. Painter in Philadelphia. Lt. Col., 116th Penn., June 26,

1862. Wd. Dec. 13, 1862, May 10, 1864 and May 31, 1864. M.O. Col., June 3, 1865. Awarded Medal of Honor for action at Battle of Chancellorsville. Chief of Police for Philadelphia. d. Philadelphia, Feb. 17, 1910.

Number: 123 **Loss:** 2 k., 11 wd., 9 m.

Monument: Sickles Ave. Map reference **V B-5**

"July 2, 1863."

118th PENNSYLVANIA INFANTRY
5th Corps, 1st Div., 1st Brig.

Other Names: "Corn Exchange Regiment"

Raised: Philadelphia, Penn.

Organized: Camp Union, near the Falls of Schuylkill, near Philadelphia, Penn. M.I. Aug. 30, 1862. Organized under the auspices of the Corn Exchange of Philadelphia.

Commander: Lt. Col. James Gwyn. b. Londonderry, Ireland, Nov. 24, 1828. Attnd. Foyle College, Ireland. Merchant in Philadelphia. Militia service. Capt. 23rd Penn. Lt. Col., 118th Penn., July 25, 1862. Wd. May 5, 1864. M.O. Col., June 1, 1865. d. Yonkers, N.Y., July 17, 1906.

Number: 332 **Loss:** 3 k., 19 wd., 3 m.

Monuments: Sickles Ave. and Big Round Top. Map references **V B-4, V I-10**

Sickles Ave—"First position July 2. July 3 on Big Round Top."

Marker: Wheatfield Road. Map reference **IV M-6**

Second position on July 2, 1863.

119th PENNSYLVANIA INFANTRY
6th Corps, 1st Div., 3rd Brig.

Other Names: "Gray Reserves"

Raised: Philadelphia, Penn. Also Delaware County.

Organized: Camp Halleck, near Philadelphia, Penn. M.I. Sept. 1, 1862.

Commander: Col. Peter Clarkson Ellmaker. b. Lancaster, Penn., Aug. 11, 1813. Officer in militia. Notary Public in Philadelphia. Col., 119th Penn., Sept. 1, 1862. M.O. Jan. 12, 1864. d. Philadelphia, Oct. 12, 1890.

Number: 466 **Loss:** 2 wd.

Monuments: Howe Ave. and Big Round Top. Map references **V K-16, V l-9**

Howe Ave—"Formed line afternoon of July 2d in rear of ridge to right of Little Round Top. Morning of the 3d moved to this position. Afternoon marched to rear of left centre. Thence to face of Round Top."

121st PENNSYLVANIA INFANTRY
1st Corps, 3rd Div., 1st Brig.
Company B, 1st Corps Provost Guard

Raised: Philadelphia. Also Venango County.

Organized: Camp "John C. Knox" near Manayunk, Philadelphia, Penn. M.I. Sept. 1, 1862.

Commanders: Maj. Alexander Biddle. b. Philadelphia, April 29, 1819. Grad. University of Pennsylvania, 1838. Militia service. Worked for counting house in Philadelphia. Maj., 121st Penn., Sept. 1, 1862. M.O. Col., Jan. 9, 1864. d. Philadelphia, May 2, 1899. Col. Chapman Biddle returned to regiment during battle to take command from Maj. Biddle. b. Philadelphia, Jan. 22, 1822. Cousin to Alexander Biddle. Grad. Saint Mary's College, Md. Lawyer in Philadelphia. Militia service. Col., 121st Penn., Sept. 1, 1862. Wd. slightly July 1, 1863. Commanded brig. June 30 and July 1. M.O. Dec. 10, 1863. d. Philadelphia, Dec. 9, 1880.

Number: 306   **Loss:** 12 k., 106 wd., 61 m.

Monuments: Reynolds Ave. and Hancock Ave. Map references **I K-4, II F-14**
Reynolds Ave—"July 1, 1863. Occupied this position the extreme left of Union line.
July 2 & 3. On Cemetery Ridge."

139th PENNSYLVANIA INFANTRY
6th Corps, 3rd Div., 3rd Brig.

Raised: Counties of Allegheny, Armstrong and Mercer.

Organized: Camp Howe near Pittsburgh, Penn. M.I. Sept. 1, 1862.

Commanders: Col. Frederick Hill Collier. b. Lancaster County, Penn., Feb. 25, 1826. Grad. Columbian College of Washington, D.C., 1849. Lawyer in Pittsburgh. Col., 139th Penn., Sept. 1, 1862. Wd. accidentally July 3, 1863. M.O. Nov. 27, 1865. d. "Rose Hill," Sharpsburg, Penn., Oct. 29, 1906. When Collier wounded, Lt. Col. William H. Moody took command of regiment. b. England. Printer in Allegheny City, Penn. Maj., 139th Penn., Sept. 1, 1862. Wd. May 12, 1864. Killed at Cold Harbor, Va., June 2, 1864, age 27 years, 2 months.

Number: 511   **Loss:** 1 k., 19 wd.

Monument: North of Wheatfield Road. Map reference **V C-10**
"Left Manchester, Md., at 9 p.m. July 1st and arrived at Rock Creek on the Baltimore Pike at 2 p.m. of the 2d. Towards evening the Brigade moved rapidly to the front to support the Union left, this Regiment deployed on the right of Little Round Top, and advanced with the 1st Brigade Penna. Reserves, driving the enemy into the Wheatfield. Retired to and held this position until the evening of the 3rd when the Regiment moved with the Penna. Reserve and advanced about 900 yards to the position indicated by a Greek Cross tablet, and assisted in forcing the enemy back. Subsequently returned to this position."

Marker: Sickles Ave. Map reference **IV L-3**
"Advanced near this point driving the enemy the evening of July 3rd."

140th PENNSYLVANIA INFANTRY
2nd Corps, 1st Div., 3rd Brig.

Raised: Counties of Washington, Beaver, Greene and Mercer.

Organized: Camp Curtin, Harrisburg, Penn. M.I. Sept. 4, 1862.
Commanders: Col. Richard Petit Roberts. b. near Frankfort Springs, Penn., June 5, 1820. Lawyer in Beaver, Penn. Capt., Co. F, 140th Penn., Aug. 26, 1862. Killed July 2, 1863. When Roberts killed, Lt. Col. John Fraser took command. b. Cromarty, Scotland, Mar. 22, 1827. Grad. University of Aberdeen, 1844. Professor at Jefferson College, Penn. A student remembered the last meeting in Fraser's classroom when the professor said: "Young gentlemen-this is our last hour of recitation together. The country needs strong and brave defenders, and since I am sound in wind and limb, I see no good reason why I should not enroll myself with them. After the exercises of Commencement Day I shall make the attempt to enlist a company from this town and its vicinity." Became Capt., Co. G, 140th Penn., Sept. 4, 1862. Wd. May 12, 1864. M.O. Col., May 31, 1865. d. Allegheny, Penn., June 4, 1878.
Number: 590 **Loss:** 37 k., 144 wd., 60 m.
Monuments: Two monuments along Sickles Ave.
Map references **V B-5, V B-6**
"The Regiment engaged the enemy on this position late in the afternoon of July 2d, succeeding 5th Corps troops and holding the right of the 1st Division, 2d Corps. Supported Battery on left centre July 3d."

141st PENNSYLVANIA INFANTRY
3rd Corps, 1st Div., 1st Brig.

Raised: Counties of Bradford, Susquehanna and Wayne.
Organized: Camp Curtin, Harrisburg, Penn. M.I. Aug. 1862.
Commander: Col. Henry John Madill. b. Hunterstown, Penn., Mar. 30, 1829. Lawyer in Towanda, Penn. Maj., 6th Penn. Reserves. Col., 141st Penn., Aug. 29, 1862. Wd. June 16, 1864 and April 2, 1865. M.O. May 28, 1865. Elected to state legislature. d. Towanda, June 29, 1899.
Number: 283 **Loss:** 25 k., 103 wd., 21 m.
Monument: Peach Orchard. Map reference **IV L-2**
"July 2 occupied this position from 4 to 6 p.m. Advanced and successfully resisted an attack on the 15th New York Light Artillery by the 2 and 8 South Carolina Infantry. Afterwards retired changed front to the right and encountered a brigade composed of the 13, 17, 18, and 21 Mississippi Infantry. Held them in check with great gallantry until outflanked. Retired firing by successive formations from the field."

142nd PENNSYLVANIA INFANTRY
1st Corps, 3rd Div., 1st Brig.

Raised: Counties of Somerset, Mercer, Westmoreland, Union, Monroe, Fayette, Venango, and Luzerne.
Organized: Camp Curtin, Harrisburg, Penn. M.I. Aug. 1862.
Commanders: Col. Robert P. Cummins. b. Somerset County, Penn., 1827. School director of Somerset borough. Capt., 10th Penn. Reserves. Resigned

Jan. 1862 having been elected sheriff of Somerset County. Col., 142nd Penn., Sept. 1, 1862. d. July 2, 1863 from wounds received on July 1. Lt. Col. Alfred Brunson McCalmont took command after Cummins. b. Franklin, Penn., April 28, 1825. Grad. Dickinson College, Penn., 1844. Lawyer in Franklin, Penn. Lt. Col., 142nd Penn., Sept. 1, 1862. M.O. Col., 208th Penn., June 1, 1865. d. Philadelphia, May 7, 1874.

Number: 362 **Loss:** 13 k., 128 wd., 70 m.

Monument: Reynolds Ave. Map reference I I-4

"July 1, a.m. Marched from near Emmittsburg [sic] reaching the field via Willoughby Run. Formed line facing northward. Occupied this position. Charged to support artillery. Reformed here and engaged a Brigade composed of the 11, 26, 47, and 52 N.C. Infantry. In the afternoon outflanked and retired firing to a position near the Seminary. Here engaged a Brigade composed of the 1, 12, 13, and 14 S.C. Infantry. After a gallant fight again outflanked and retired to Cemetery Hill.

July 2. In position at Cemetery Hill.

July 3. Moved half a mile to the left and exposed to the artillery fire of the enemy."

143rd PENNSYLVANIA INFANTRY
1st Corps, 3rd Div., 2nd Brig.

Raised: Counties of Luzerne and Susquehanna.

Organized: Camp Luzerne, Kingston Township, near Wilkes Barre, Penn. M.I. Aug.-Sept., 1862.

Commanders: Col. Edmund Lovell Dana. b. Wilkes Barre, Penn., Jan. 29, 1817. Grad. Yale College, 1838. Mexican War service. Officer in militia. Lawyer in Wilkes Barre. Col., 143rd Penn., Nov. 18, 1862. Wd. Aug. 3, 1864. M.O. Aug. 4, 1865. d. Wilkes Barre, April 25, 1889. When Dana promoted to brigade command, Lt. Col. John Dunn Musser commanded regiment. b. Lewisburg, Penn., April 24, 1826. Made two trips to California gold fields in 1849 and 1851. Residence Lewisburg, Penn. at the start of the war. Lt., Co. K, 143rd Penn., Oct. 1, 1862. Killed in the Battle of the Wilderness, Va., May 6, 1864.

Number: 515 **Loss:** 21 k., 141 wd., 91 m.

Monuments: Reynolds Ave. and Hancock Ave. Map references I G-4, II E-15
Reynolds Ave—"This monument marks right of first position July 1, 1863, facing north and second position facing west, which the Regiment held from 11:30 a.m. until First Corps fell back. Last position on Seminary Ridge right resting on Railroad Cut.

July 2 and 3. Regiment was in line on left centre and on the 3d assisted in repulsing the final charge of the enemy."

145th PENNSYLVANIA INFANTRY
2nd Corps, 1st Div., 4th Brig.

Raised: Counties of Erie, Warren, Crawford and Mercer.

Organized: Camp on fairgrounds near Erie, Penn. M.I. Sept. 5, 1862. Most of Companies C, H, and K, were captured at Chancellorsville, Va., May 1863.

Commanders: Col. Hiram Loomis Brown. b. North East, Penn., Oct. 27, 1832. Hotel keeper in Erie, Penn. Militia officer. Capt., "Erie Regt." (3 mos.). Capt., 83rd Penn. Infantry. Col., 145th Penn., Sept. 5, 1862. Wd. July 27, 1862 and July 2, 1863. M.O. Feb. 1, 1865. Sheriff of Erie County. d. Erie, Nov. 25, 1880. When Brown wd. at Gettysburg, Capt. John William Reynolds took command. b. Evansburg, Penn., July 3, 1836. Freight agent for railroad in Erie, Penn. Capt., Co. A, 145th Penn., Aug. 26, 1862. Wd. July 2, 1863. M.O. Maj., Sept. 19, 1863. d. Erie, Oct. 24, 1925. Third commander at Gettysburg was Capt. Moses Warren Oliver. b. South Danville, N.Y., June 8, 1833. Instructor at State Normal School in Edinboro, Penn. Capt., Co. B, 145th Penn., Aug. 26, 1862. M.O. Nov. 24, 1863. Elected to state legislature. d. Conneautville, Penn., Feb. 4, 1906.

Number: 228 **Loss:** 11 k., 69 wd., 10 m.

Monument: Brooke Ave. Map reference V C-4

"July 2. In the evening about 5 O'Clock the Regiment with the Brigade charged from the northerly side of the Wheatfield driving the enemy and capturing many prisoners. This position was held until the command was outflanked when it retired under orders.

July 3. The Regiment was in position on the left center with the Division."

147th PENNSYLVANIA INFANTRY
8 Companies
12th Corps, 2nd Div., 1st Brig.

Raised: Philadelphia. Counties of Dauphin, Allegheny, Luzerne and Huntingdon.

Organized: Harpers Ferry, West Virginia. M.I. Oct. 1862. Formed by combining companies LMNOP of the 28th Penn. Infantry with three new companies from Harrisburg, Penn. Companies I and K were not formed until after the Battle of Gettysburg.

Commander: Lt. Col. Ario Pardee, Jr. b. Hazleton, Penn., Oct. 28, 1839. Grad. Rensselaer Polytechnic Institute, N.Y., 1858. Civil Engineer in charge of father's coal mines. Militia officer. Capt., 28th Penn. Inf. Lt. Col., 147th Penn., Oct. 10, 1862. M.O. Col., June 13, 1865. d. Wyncote, Penn., Mar. 16, 1901.

Number: 298 **Loss:** 5 k., 15 wd.

Monument: Geary Ave. Map reference III E-7

"On the night of July 1st this Regiment lay on the northern slope of Little Round top holding the extreme left of the Union Army. At 6 p.m. July 2d moved to Culp's Hill where it was held in reserve until evening; then marched toward the left with the Brigade returning at about 3 a.m. July 3d and occupied this position."

Markers: Company G near monument, tablet on boulder near Geary Ave., and along Sykes Ave. Map references III E-7, III F-7, V D-12

148th PENNSYLVANIA INFANTRY
2nd Corps, 1st Div., 1st Brig.

Raised: Counties of Centre, Jefferson and Clarion.

Organized: Camp Curtin, Harrisburg, Penn. M.I. Sept. 1, 1862.

Commanders: Col. Henry Boyd McKeen. b. Philadelphia, Sept. 18, 1835. Grad. Princeton College, N.J., 1853. Lumber merchant living in Camden, N.J. Lt. and Adjt., 81st Penn. Inf., Oct. 27, 1861. Wd. Dec. 1862. Killed June 3, 1864 at Cold Harbor, Va. McKeen given temporary command of 148th Penn. on June 30, 1863. When he moved to brigade command, Lt. Col. Robert McFarlane led regiment. b. on a farm near Boalsburg, Penn., Nov. 6, 1826. Lived there at the start of the war. Capt., 7th Penn. Infantry (3 mos.). Capt., Co. G, 148th Penn., Aug. 27, 1862. M.O. Nov. 4, 1863 for disability. d. Bellefonte, Penn., May 18, 1891.

Number: 468 **Loss:** 19 k., 101 wd., 5 m.

Monuments: Ayres Ave. and Hancock Ave. Map references **V C-8, IV H-13**
Ayres Ave—"The Regiment engaged the enemy on this position in the afternoon of July 2, 1863."

149th PENNSYLVANIA INFANTRY
1st Corps, 3rd Div., 2nd Brig.
Company D, Provost Guard, 3rd Div., 1st Corps

Other Names: "Second Bucktails"

Raised: Counties of Potter, Tioga, Clearfield, Lebanon, Mifflin and Huntingdon.

Organized: In response to the notoriety received by the 1st Bucktail regiment (13th Penn. Reserves), a Bucktail Brigade was authorized. Only two regiments, the 149th Penn. and 150th Penn., were formed when they were ordered to the Army of the Potomac for the Antietam campaign. The 149th was organized at Camp McNeil, Harrisburg, Penn. M.I. Aug. 30, 1862.

Commanders: Col. Walton Dwight. b. Windsor, N.Y., Dec. 20, 1837. Lumber business in Coudersport, Penn. Capt., Co. K, 149th Penn., Aug. 27, 1862. Wd. July 1, 1863. M.O. Mar. 31, 1864. Mayor of Binghamton, N.Y., 1871-1873. d. Binghamton, Nov. 15, 1878. When Dwight wounded, Capt. James Glenn took command. b. Allegheny County, July 23, 1824. Militia service. Businessman in Allegheny County at start of war. Capt., Co. D, 149th Penn., Aug. 23, 1862. M.O. Lt. Col., Aug. 4, 1865. Feed and grain business in Pittsburgh. d. Aug. 23, 1901 at Glenn's Station, near Carnegie, on part of the old Glenn family homestead where he was born.

Number: 450 **Loss:** 53 k., 172 wd., 111 m.

Monument: Chambersburg Pike. Map reference **I G-4**
"July 1. The Regiment held this position from 11:30 a.m. until the Corps retired, resisting several assaults of the enemy, making two successful charges to the R.R. Cut and changing front to rear under fire.
July 2. Moved to support of the left and remained on picket all night. In the morning of the 3d moved to left centre where its other monument stands. "

Markers: Hagerstown Road (Company D) and Hancock Ave. Map references I K-7, II F-15

Hagerstown Road—"Held ground for 20 minutes on evening of July 1."

150th PENNSYLVANIA INFANTRY
9 Companies
1st Corps, 3rd Div., 2nd Brig.

Other Names: "Third Bucktails"

Raised: Philadelphia. Counties of Crawford, Union and McKean.

Organized: Camp Curtin, Harrisburg, Penn. M.I. Sept. 4, 1862. See 149th Penn. for story of "Bucktail Brigade." Company D permanently assigned to guard President Lincoln.

Commanders: Col. Langhorne Wister. b. "Belfield" near Germantown, Philadelphia, Sept. 20, 1834. Iron business in Duncannon, Penn. Capt., 13th Penn. Reserves. Col. 150th Penn., Sept. 5, 1862. Wd. June 27, 1862 and July 1, 1863. M.O. Feb. 22, 1864. d. at the home where he was born on Mar. 19, 1891. When Wister promoted to brigade command, Lt. Col. Henry Shippen Huidekoper took over regiment. b. "Pomona," Meadville, Penn., July 17, 1839. Grad. Harvard University, 1862. Capt., Co. K, 150th Penn., Sept. 5, 1862. Wd. July 1, 1863. Right arm amputated in Gettysburg Catholic Church. Awarded Medal of Honor for gallantry at Gettysburg. M.O. Mar. 5, 1864. Active in National Guard. d. Philadelphia, Nov. 9, 1918. Capt. George W. Jones took command on night of July 1, 1863. b. Philadelphia, Nov. 4, 1833. Carpenter. Sergt., 22nd Penn. Infantry (3 mos.). Sergt., Phelps Missouri Infantry (6 mos.). Capt., Co. B, 150th Penn., Aug. 25, 1862. M.O. Col., June 23, 1865. d. Germantown, Philadelphia, Nov. 26, 1913.

Number: 397 **Loss:** 35 k., 152 wd., 77 m.

Monuments: Stone Ave. and Hancock Ave. Map references I G-3, II F-14

Stone Ave—"July 1 the Regiment held this position from 11:30 a.m. to 3:30 p.m. This monument marks the most advanced line, facing west, occupied by the Regiment. Repeated changes of front were made to meet assaults from the north and west, and the right wing charged to R.R. cut. In retiring it made several stands and engaged the enemy. Evening of the 2d moved to support the left, and held position on Emmittsburg [sic] Road. Morning of the 3d moved to the left centre and remained until close of the battle."

151st PENNSYLVANIA INFANTRY
1st Corps, 3rd Div., 1st Brig.

Raised: Counties of Berks, Susquehanna, Pike, Warren and Juniata.

Organized: 9 months regiment formed at Camp Curtin, Harrisburg, Penn. M.I. Sept.-Oct., 1862.

Commanders: Lt. Col. George Fisher McFarland. b. near Harrisburg, Penn., April 28, 1834. Headed a school at McAlisterville, Penn., from which he drew recruits for a military company. Became Capt., Co. D, 151st Penn.,

Oct. 24, 1862. Wd. July 1, 1863. right leg amputated. M.O. July 27, 1863. Instrumental in establishing a state system of orphans' schools. d. Tallapoosa, Ga., Dec. 18, 1891. When McFarland wd., Capt. Walter L. Owens commanded regiment. b. Mifflin County, Penn., around 1840. Teacher in McFarland's school. Sergt., Co. D, 151st Penn., Nov. 19, 1862. M.O. July 27, 1863. Lived in Granville, Penn. d. April 10, 1912.

Number: 467 **Loss:** 51 k., 211 wd., 75 m.

Monument: Reynolds Ave. Map reference I I-4

"July 1. Fought here and in the Grove west of the Theological Seminary. July 2. In reserve on Cemetery Hill. July 3. In position on left centre and assisted in repulsing the charge of the enemy in the afternoon."

153rd PENNSYLVANIA INFANTRY
11th Corps, 1st Div., 1st Brig.

Raised: Northampton County

Organized: 9 months regiment formed at Camp Curtin, Harrisburg, Penn. M.I. Oct. 7, 1862.

Commander: Maj. John Frederick Frueauff. b. Nazareth, Penn., May 25, 1838. Militia service. Lawyer in Bethlehem, Penn. Lt., 1st Penn. Infantry (3 mos.). Maj., 153rd Penn., Oct. 11, 1862. Wd., May 3, 1863. M.O. July 24, 1863. d. Leadville, Colorado, Nov. 8, 1886.

Number: 569 **Loss:** 23 k., 142 wd., 46 m.

Monument: Howard Ave. Map reference I B-15

"July 1. The Regiment held this position in the afternoon until the Corps was outflanked and retired, when it took position along the lane at the foot of East Cemetery Hill, where it remained until the close of the battle, assisting to repulse the enemy's assault on the night of the 2d."

Marker: Wainwright Ave. Map reference II L-4

155th PENNSYLVANIA INFANTRY
5th Corps, 2nd Div., 3rd Brig.

Raised: Counties of Allegheny, Clarion and Armstrong.

Organized: Camp Howe, Pittsburgh, Penn. M.I. Sept. 2, 1862.

Commander: Lt. Col. John Herron Cain. b. Pittsburgh, Penn., Nov. 18, 1838. Bank teller in Chattanooga, Tenn. Returned to Pittsburgh to enlist. Pvt. 12th Penn. Infantry (3 mos.). Capt., Co. C, 155th Penn., Aug. 29, 1862. M.O. Col., Aug. 30, 1863. d. Franklin, Penn., April 28, 1903.

Number: 424 **Loss:** 6 k., 13 wd.

Monument: Sykes Ave. Map reference V E-11

"Position occupied July 2nd, 3rd, & 4th, 1863."

1st PENNSYLVANIA ARTILLERY, BATTERY B
1st Corps Artillery Brigade

Other Names: "43rd Pennsylvania Volunteers"

Raised: Lawrence County

Organized: Mount Jackson, Penn. M.I. Aug. 5, 1861.

Commander: Capt. James Harvey Cooper. b. Rops Township, Penn., Mar. 8, 1840. Clerk in Mt. Jackson. Capt., Battery B, Aug. 5, 1861. M.O. Maj., Aug. 9, 1864. Merchant in New Castle, Penn. d. Mar. 21, 1906.

Number: 4 Ordnance Rifles. 114 men. **Loss:** 3 k., 9 wd.

Monuments: Reynolds Ave. and two on East Cemetery Hill. Map references I I-4, II K-4

Reynolds Ave—"July 1, 1863 the Battery arrived here about noon and engaged Confederate Artillery on Herr's Ridge. About 1:30 p.m. moved to the rear. Changed front, engaged Carter's Artillery and shelled Rodes' Infantry on Oak Hill. About 3 p.m. moved to the woods in front of Theological Seminary and resisted the final attack of Scales', Perrin's and other Brigades."

East Cemetery Hill—"July 1, 1863: Battery arrived at 12 m. took position and was engaged between Hagerstown Road and Chambersburg Pike near Willoughby Run; changed position to the right and swept Oak Hill with its fire. Withdrew to Theological Seminary, where it fought till after 4 p.m.; retired to this position where it remained until close of heavy artillery contest with the enemy's batteries on Benner's Hill, during afternoon engagement of July 2, when relieved by Ricketts' Battery. July 3: Was engaged on left center during the final attack and second repulse of the enemy."

Marker: Hancock Ave. Map reference IV G-13

Position of July 3, 1863.

1st PENNSYLVANIA ARTILLERY,
BATTERIES F&G TEMPORARILY CONSOLIDATED
Artillery Reserve, 3rd Volunteer Brigade

Raised: G-Philadelphia, Penn. F-Schuylkill County.

Organized: Camp Curtin, Harrisburg, Penn. M.I. Aug. 5, 1861.

Commander: Capt. Robert Bruce Ricketts. b. Orangeville, Penn., April 29, 1839. Law student when war began. Lt., Battery F, Aug. 5, 1861. M.O. Maj., June 5, 1865. Lawyer in Wilkes Barre, Penn. d. North Mountain, Penn., Nov. 13, 1918.

Number: 6 Ordnance Rifles. 144 men. **Loss:** 6 k., 14 wd., 3 m.

Monument: East Cemetery Hill. Map reference II K-3

"July 2nd. Reached the field and took this position in the afternoon and engaged the Rebel batteries on Benner's Hill. 8 p.m. A Rebel column charged the Battery and a desperate hand-to-hand conflict ensued which was repulsed after every round of canister had been fired.

July 3rd. Engaged with the Rebel batteries on the left and centre of the line."

PENNSYLVANIA INDEPENDENT BATTERIES
C&F TEMPORARILY CONSOLIDATED
Artillery Reserve, 1st Volunteer Brigade

Raised: Allegheny County

Organized: Battery C-Pittsburgh, Penn. Battery F, Camp Lamon, Williamsport,

Maryland. M.I. Oct.-Nov., 1861. From June 3, 1863 to March 25, 1864, Batteries F and C served as a consolidated battery.

Commander: Capt. James Thompson. b. County Down, Ireland, 1821. British artillery service. Came to U.S. in 1856. Painter in Allegheny City. Capt., Battery C, Sept. 24, 1861. M.O. June 30, 1865. d. at daughter's residence in Ingram, Penn., Mar. 1906.

Number: 6 Ordnance Rifles. 105 men.   **Loss:** 2 k., 23 wd., 3 m.

Monument: Peach Orchard. Map reference **IV K-1**

Peach Orchard

Battery C. "July 2. Occupied this position from about 5 to 6 O'Clock p.m. July 3. In position on right of First Volunteer Brigade Reserve Artillery and engaged the enemy."

Battery F. "July 2. Occupied this position from about 5 to 6 O'Clock p.m. July 3. With the left centre on Cemetery Ridge on left of First Volunteer Brigade Reserve Artillery marked by tablet. 24 men from Battery F were detailed to Battery H, 1st Ohio Artillery posted in the cemetery during the battle."

Marker: Hancock Ave. Map reference **IV E-12**

PENNSYLVANIA INDEPENDENT BATTERY E
12th Corps Artillery Brigade

Other Names: "Knap's"

Raised: Philadelphia, also Allegheny County.

Organized: Camp De Korponay, Point of Rocks, Maryland. M.I. Sept. 1, 1861.

Commander: Lt. Charles A. Atwell. b. Pittsburgh, Penn., 1840 or 1841. Clerk in Allegheny City, Penn. Pvt., 12th Penn. Inf. (3 mos.). Lt., Battery E, Sept, 21, 1861. d. Nov. 2, 1863 from wounds received Oct. 28, 1863 at Wauhatchie, Tenn. when capt. of battery.

Number: 6 10-pdr. Parrotts. 139 men.   **Loss:** 3 wd.

Monuments: Powers Hill and Culp's Hill. Map references **III E-18, III F-3**

Powers Hill—"At 3:30 p.m. July 2nd one gun was placed on Culp's Hill in the position marked by a monument, and was joined by two others at 5 p.m., when the three guns engaged the enemy's batteries on Benner's Hill. These guns were withdrawn when the Infantry was ordered to the left and the Battery went into this position where it remained until the close of the battle."

3rd PENNSYLVANIA HEAVY ARTILLERY, BATTERY H
Attached to Cavalry Corps, 2nd Div., 1st Brig.

Other Names: "152nd Pennsylvania Volunteers"

Raised: Lebanon County.

Organized: Camp Ruff, Camden, N.J. M.I. Sept. 1862. Co. H ordered to duty in the defenses of Baltimore, Md. where it remained for term of service except for the Gettysburg campaign.

Commander: Capt. William D. Rank. b. Philadelphia, July 12, 1838. Clerk in Philadelphia. Capt., Battery H, Jan. 19, 1863. M.O. July 25, 1865. Hardware merchant in Philadelphia. d. Philadelphia, Jan. 17, 1872.
Number: 2 Ordnance Rifles. 52 men.   **Loss:** 1 m.
Monument: Hanover Road. No map reference.
Marker: Hancock Ave. Position of July 3. Map reference **IV H-13**

1st PENNSYLVANIA CAVALRY
9 Companies, Army Headquarters
Company H, 6th Corps Headquarters

Other Names: "15th Reserve." "44th Pennsylvania Volunteers."
Raised: Counties of Juniata, Montgomery, Mifflin, Clinton, Centre, Greene, Fayette, Washington, Allegheny and Berks.
Organized: Camp Jones, near Washington, D.C. M.I. Aug. 28, 1861. Originally formed to be part of the Pennsylvania Reserves. Companies G and L remained at Frederick, Md. during the Gettysburg campaign.
Commander: Col. John P. Taylor. b. Reedsville, Penn., June 6, 1827. Farmer in Reedsville. Militia service. Capt., Company C, 1st Penn. Cav., Aug. 27, 1861. Wd. Aug. 9, 1862. M.O. Sept. 9, 1864. d. Reedsville, Penn., June 27, 1914. Coffin said to be made of melted down Civil War cannon.
Number: 344   **Loss:** 2 m.
Monument: Hancock Ave. Map reference **II F-12**
"At the opening of the artillery fire on the afternoon of July 3 the Regiment was in line to the left and rear of this position with orders from General Meade to 'charge the assaulting column should it succeed in breaking the infantry line in front.'"

2nd PENNSYLVANIA CAVALRY
Army Headquarters

Other Names: "59th Pennsylvania Volunteers"
Raised: Philadelphia. Counties of Lancaster, Centre, Crawford, Tioga, Armstrong and Northampton.
Organized: Camp Patterson, Point Breeze, near Philadelphia, Penn. M.I. Dec. 1861-Jan. 1862.
Commander: Col. Richard Butler Price. b. Philadelphia, Dec. 15, 1807. Militia service. Studied in France. Served for 3 months on General Robert Patterson's staff. Col., 2nd Penn. Cav., Jan. 23, 1862. M.O. Feb. 1, 1865. d. Philadelphia, July 15, 1876.
Number: 575   **Loss:** None
Monument: Meade's Headquarters. Map reference **II H-10**
"The Regiment held this position July 3 until the close of the day when it conducted 3000 prisoners to Westminster, Md. Detachments served on other parts of the field during the battle."

139

3rd PENNSYLVANIA CAVALRY
Cavalry Corps, 2nd Div., 1st Brig.

Other Names: "Kentucky Light Cavalry." "60th Pennsylvania Volunteers."
Raised: Philadelphia, Penn., Washington, D.C. and Penn. counties of Allegheny, Cumberland and Schuylkill.
Organized: Camp Park, Washington, D.C. M.I. July and Aug., 1861. Regiment formed by Col. William H. Young from Kentucky.
Commander: Lt. Col. Edward S. Jones. b. Philadelphia, Penn., Aug. 4, 1818. Publisher living in Germantown. Capt., Company C, 3rd Penn. Cavalry, Aug. 1, 1861. M.O. Aug. 24, 1864. d. Nashville, Tennessee, Nov. 25, 1886. Buried National Cemetery there.
Number: 394   **Loss:** 15 wd., 6 m.
Monument: Gregg Ave. East Cavalry Battlefield. No map reference.
"July 2nd 1863. Reached the field at noon from Hanover. Engaged dismounted a Confederate Brigade of Infantry on Brinkerhoff's Ridge from 6 to 10 p.m. July 3rd. Engaged mounted and dismounted with the Confederate Cavalry Division on this field from 2 p.m. until evening, portions of the Regiment advancing in a mounted charge and driving the enemy beyond the Rummel Farm Buildings."
Marker: East Cavalry Battlefield along Low Dutch Road. No map reference.

4th PENNSYLVANIA CAVALRY
Cavalry Corps, 2nd Div., 3rd Brig.

Other Names: "64th Pennsylvania Volunteers"
Raised: Counties of Venango, Allegheny, Westmoreland, Lebanon and Luzerne.
Organized: Camp Curtin, Harrisburg, Penn. M.I. Aug. 15 to Oct. 30, 1861.
Commander: Lt. Col. William Emile Doster. b. Bethlehem, Penn., Jan. 8, 1837. Grad. Yale College, 1857. Attnd. Harvard University Law School and the University of Heidelberg, Germany. Law student in Philadelphia at start of war. Capt., Harlan's Light Cavalry. Maj., 4th Penn. Cav., Oct. 28, 1861. M.O. Dec. 7, 1863. Appointed to defend two of the Lincoln conspirators in the assassination trial. d. Bethlehem, Penn., July 2, 1919.
Number: 307   **Loss:** 1 k.
Monument: Hancock Ave. Map reference **IV H-14**
"Detached on the morning of July 2nd from the Brigade at the junction of White Run and Baltimore Turnpike, ordered to report to Headquarters, Army of the Potomac. Supported a battery temporarily near this position. On picket at night, retiring late on the afternoon of the 3rd to Second Cavalry Division."

6th PENNSYLVANIA CAVALRY
9 Companies, Cavalry Corps, 1st Div., Reserve Brigade
Companies E and I at Army Headquarters

Other Names: "Rush's Lancers." "70th Pennsylvania Volunteers."

Raised: Philadelphia. Also Berks County.

Organized: Camp Meigs, Philadelphia, Penn. M.I. Oct. 31, 1861. Distinctive lances carried by the regiment at the start of the war were turned in May 24, 1863. Companies A, D and F on detached service during the Gettysburg campaign.

Commander: Maj. James Henry Haseltine. b. Philadelphia, Penn., Nov. 2, 1833. Brother was artist W. S. Haseltine. J. H. was a sculptor trained in Paris and Rome. He wrote that he had "returned to this country on a visit and war breaking out I thought it was my duty to enter the service ...." Pvt., Philadelphia City Troop (3 mos.). Capt., Co. E, 6th Penn. Cavalry, Sept. 18, 1861. M.O. Nov. 12, 1863. Returned to Europe. d. Rome, Italy, Nov. 9, 1907.

Number: 366 **Loss:** 3 k., 7 wd., 2 m.

Monuments: Emmitsburg Road and near Meade's Headquarters (E&I). Map reference **II H-10**

Emmitsburg Road monument shown on key map. "This Regiment detached with the 2nd Corps covered the rear of the army on the march from Virginia. At Frederick rejoined the Cavalry Corps and with Gregg's Division moved in advance to Gettysburg. July 1st moved hastily to Manchester to protect trains. July 4th joined in pursuit of the enemy participating in the night attack on Monterey Pass...."

8th PENNSYLVANIA CAVALRY

Detached with its brigade in Maryland during the Battle of Gettysburg.

16th PENNSYLVANIA CAVALRY
Cavalry Corps, 2nd Div., 3rd Brig.

Other Names: "161st Pennsylvania Volunteers"

Raised: Philadelphia. Counties of Venango, Fayette, Erie, Juniata, Franklin and Washington.

Organized: Camp Simmons, Harrisburg, Penn. M.I. Sept.-Nov., 1862.

Commander: Lt. Col. John Kincaid Robison. b. Milford Township, Juniata County, Penn., July 17, 1829. Farmer in Patterson, Penn. Capt., 1st Penn. Cavalry. Capt., Co. F, 16th Penn. Cavalry, Oct. 10, 1862. Wd. Oct. 12, 1863 and April 7, 1865. M.O. Aug. 11, 1865. d. Mifflintown, Penn., June 20, 1917.

Number: 411 **Loss:** 2 k., 4 wd.

Monument: Highland Avenue Road. See Key Map near Wolf Hill. "Position occupied on the afternoon of July 3d, 1863."

17th PENNSYLVANIA CAVALRY
Cavalry Corps, 1st Div., 2nd Brig.
Companies DH, 5th Corps Headquarters
Company K, 11th Corps Headquarters

Other Names: "162nd Pennsylvania Volunteers"

141

Raised: Counties of Beaver, Susquehanna, Lancaster, Bradford, Lebanon, Cumberland, Franklin, Schuylkill, Luzerne and Wayne.

Organized: Camp Simmons, Harrisburg, Penn. M.I. Oct. 18, 1862.

Commander: Col. Josiah Holcomb Kellogg. b. Erie, Penn., Oct. 1, 1836. Attnd. Hobart College, N.Y. Grad. West Point, 1860. 1st U.S. Cavalry. Col., 17th Penn. Cav., Nov. 19, 1862. M.O. Dec. 27, 1864. Retd. Feb. 6, 1865. d. Chicago, Ill., June 19, 1919.

Number: 448   **Loss:** 4 m.

Monument: Buford Ave. Map reference **I B-6**

"The Regiment held this position on the morning of July 1, 1863, from 5 O'Clock until the arrival of First Corps troops. The Brigade then moved to the right, covering the roads to Carlisle and Harrisburg and holding the enemy in check until relieved by troops of the Eleventh Corps. It then took position on the right flank of the infantry, and, later, aided in covering the retreat of the 11th Corps to Cemetery Hill, where it went into position with the Division on the left of the army."

18th PENNSYLVANIA CAVALRY
Cavalry Corps, 3rd Div., 1st Brig.

Other Names: "163rd Pennsylvania Volunteers"

Raised: Philadelphia. Counties of Greene, Crawford, Dauphin, Washington, Allegheny, Lycoming and Cambria.

Organized: Camp Curtin, Harrisburg, Penn. M.I. Aug.-Nov., 1862.

Commander: Lt. Col. William Penn Brinton. b. Paradise Township, Penn., May 15,1832. From a prominent Lancaster County family. Probably lived in Philadelphia at start of the war. Capt., 2nd Penn. Cavalry. Lt. Col., 18th Penn., May 2, 1863. M.O. Jan. 13, 1865. In 1879 he sailed for South America as agent for several business concerns. Around 1881 he went into the interior of Argentina as a school teacher and never returned.

Number: 599   **Loss:** 2 k., 4 m., 8 wd.

Monument: Near Confederate Ave. Map reference **V K-3**

"The Regiment participated in the cavalry fights at Hanover June 30th and Hunterstown July 2d 1863. On July 3d occupied this position, and in the afternoon charged with the Brigade upon the enemy's infantry behind the stone wall to the north of this point on the outer edge of the woods."

2nd RHODE ISLAND INFANTRY
6th Corps, 3rd Div., 2nd Brig.

Raised: From every county in the state, especially Providence, Kent and Washington.

Organized: Camp Burnside on Dexter Training Ground in Providence, R.I. M.I. June 1861.

Commander: Col. Horatio Rogers, Jr. b. Providence, R.I. May 18, 1836. Attnd. Harvard University Law School. Grad. Brown University, R.I., 1855. Lawyer in Providence. June 1861, elected Justice of City Police Court. Lt., 3rd R.I. Artillery. Col., 11th R.I. Inf. (9 mos.). Col., 2nd R.I. Infantry, Feb. 6, 1863. M.O. Jan. 15, 1864. Of the carnage at Gettysburg, Rogers wrote, "Death seemed to be holding a carnival." Member of the state legislature and Attorney General of Rhode Island. d. Providence, R.I., Nov. 12, 1904.

Number: 409   **Loss:** 1 k., 5 wd., 1 m.

Monument: Sedgwick Ave. Map reference **V C-13**
"July 2 & 3, 1863."

Marker: Emmitsburg Road. Map reference **IV B-7**
Marks the skirmish line on July 4, 1863.

1st RHODE ISLAND ARTILLERY, BATTERY A
2nd Corps Artillery Brigade

Raised: Providence County

Organized: Camp Burnside, Providence, R.I. M.I. June 6, 1861.

Commander: Capt. William Albert Arnold. b. Cranston, R.I., Sept. 4, 1830. Bookkeeper in Providence. 2 Lt., 1st R.I. Artillery, Sept. 1861. M.O. June 17, 1864. d. soldier's home in Bath, New York, Mar. 24, 1908.

Number: 6 Ordnance Rifles. 139 men.   **Loss:** 3 k., 28 wd., 1 m.

Monument: Hancock Ave. Map reference **II E-11**
"July 2 & 3, 1863."

1st RHODE ISLAND ARTILLERY, BATTERY B
2nd Corps Artillery Brigade

Other Names: "Hazard's"

Raised: Providence County

Organized: Providence, R.I. Armory. M.I. Aug. 13, 1861. Assigned to the 2nd Corps Artillery Brigade from the Artillery Reserve on July 1, 1863.

Commanders: Lt. Thomas Frederic Brown. b. Providence, R.I., Oct. 26, 1842. Start of war was in his third year at Brown University, R.I. 5' tall. Corpl., 1st R.I. Artillery, June 6, 1861. Lt., Battery B, Dec. 29, 1862. Wd. July 2, 1863. M.O. Capt., June 12, 1865. Lived in Cincinnati, Oh. at least until 1898. d. Daytona, Florida, Nov. 27, 1928. When Brown wounded, Lt. William Smith Perrin took command of battery. b. Nov. 12, 1839. Lived in Smithfield, R.I. at start of the war. Corpl., 1st R.I. Artillery, Aug. 25, 1861. 2 Lt., Battery B, Nov. 11, 1862. Wd. Aug. 25, 1864, requiring amputation of right leg. M.O. Feb. 4, 1865. Clerk in Lime Rock, R.I. d. Pawtucket, R.I., Aug. 13, 1876, from an overdose of morphine.

Number: 6 12-pdr. Napoleons. 103 men.   **Loss:** 7 k., 19 wd., 2 m.

Monument: Hancock Ave. Map reference **II F-13**

Marker: Field west of copse of trees, Hancock Ave. Map reference **II D-13**

1st RHODE ISLAND ARTILLERY, BATTERY C
6th Corps Artillery Brigade

Raised: Providence County

Organized: Camp Ames, Providence, R.I. M.I. Aug. 25, 1861.

Commander: Capt., Richard Waterman. b. Providence, R.I., Jan. 29, 1839. Grad. Brown University, R.I., 1859. Start of war ceased mercantile pursuits in New York and returned to Providence to join militia battery. Lt., Battery C, Aug. 25, 1861. M.O. Sept. 2, 1864. d. Providence, Mar. 23, 1888.

Number: 6 10-pdr. Parrotts. 125 men.   **Loss:** None

Monument: None.

1st RHODE ISLAND ARTILLERY, BATTERY E
3rd Corps Artillery Brigade

Other Names: "Randolph's"

Raised: Providence County.

Organized: Camp Greene, near Providence, R.I. M.I. Sept. 30, 1861.

Commanders: Lt. John Knight Bucklyn. b. Foster, R.I., Mar. 15, 1834. Grad. Brown University, R.I., 1861. School teacher. Sergt., Battery E, Sept. 30, 1861. Wd. July 2, 1863. M.O. Feb. 2, 1865. Awarded Medal of Honor for gallantry at the Battle of Chancellorsville, Va. d. Mystic, Conn., May 15, 1906. Next in line for command at Gettysburg was 2 Lt. Benjamin Freeborn. b. Providence, R.I., Jan. 3, 1835. Cashier for transportation company in St. Louis, Missouri. Returned to Rhode Island to enlist. Pvt., 1st R.I. Artillery, Dec. 2, 1861. Joined Battery E, April 20, 1863. Slightly wounded July 2, 1863. M.O. Lt., June 24, 1865. Returned to St. Louis. d. May 20, 1874 the day after he fell from his carriage, striking his head.

Number: 6 12-pdr. Napoleons. 116 men.   **Loss:** 3 k., 26 wd., 1 m.

Monument: Emmitsburg Road. Map reference **IV J-1**
   "July 2, 1863."

1st RHODE ISLAND ARTILLERY, BATTERY G
6th Corps Artillery Brigade

Raised: Providence County.

Organized: Providence, R.I. M.I. Dec. 1861. See entry for 10th N.Y. Ind. Battery.

Commander: Capt., George William Adams. b. Providence, R.I., Oct. 15, 1834. Attnd. Brown University, R.I. Bookseller in Providence. Lt., 1st R.I. Artillery, Aug. 13, 1861. Capt., Battery G, Jan. 30, 1863. M.O. June 1865. d. Bristol, R.I., Oct. 13, 1883.

Number: 6 Ordnance Rifles. 135 men.   **Loss:** None

Monument: None

78th and 102nd New York Infantry, North Slocum Avenue.

145

2nd UNITED STATES INFANTRY
6 Companies, BCFHIK
5th Corps, 2nd Div., 2nd Brig.

Raised: Enlisted at New York City; Suffolk County, Mass.; and St. Louis County, Missouri.

Organized: 1815. On April 12, 1861, headquarters of the regiment was at Fort Kearny, Nebraska Territory.

Commanders: Maj. Arthur Tracy Lee. b. Northumberland, Penn., June 26, 1814. 2 Lt., 5th U.S. Inf., Oct. 8, 1838. Maj., 2nd U.S. Inf., Oct. 26, 1861. Wd. July 2, 1863. Retd. with rank of Col., July 28, 1866. d. Rochester, N.Y., Dec., 29, 1879. When Lee wd., Capt. Samuel A. McKee, Jr. led the regiment. b. Birmingham, Penn., Oct. 8, 1841. Attnd. Western Military Institute in Kentucky. Returned to Birmingham to join Co. B, 62nd Penn. as Lt. on July 22, 1861. 2 Lt., 2nd U.S. Inf., Aug. 5, 1861. Wd. Sept. 17, 1862. Killed April 11, 1864 by guerrillas near Greenwich, Va.

Number: 237 **Loss:** 6 k., 55 wd., 6 m.

Monument: Ayres Ave. Map reference **V C-8**

"July 2. Arrived in the morning and took position with the Brigade at the right of the Twelfth Corps. Skirmished with the Confederates. Later moved to the left. At 5 p.m. formed line with left on the north slope of Little Round Top and the right of Brigade line extending into some woods. Advanced across Plum Run and to the crest of the rocky wooded hill in front near the Wheatfield and, facing left, occupied the stone wall on the edge of the woods. The Confederates, having opened fire on the right flank and advanced through the Wheatfield in the rear, the Brigade was withdrawn under a heavy infantry fire on both flanks and from the rear and of shot and shell from the Batteries, and formed in line on right of Little Round Top.

July 3. Remained in same position."

3rd UNITED STATES INFANTRY
6 Companies, BCFGIK
5th Corps, 2nd Div., 1st Brig.

Other Names: "Buff Sticks"

Raised: Enlisted at New York City and Suffolk County, Mass.

Organized: 1815. On April 12, 1861, the regiment left Indianola, Texas by ship bound for New York Harbor.

Commanders: Capt. Henry William Freedley. b. Penn. Family from Norristown, Penn. Entered West Point from Norristown in 1851. 2 Lt., 9th U.S. Inf., 1855. 2 Lt., 3rd U.S. Inf., Sept. 30, 1855. Wd. July 2, 1863. Retd. with rank of Col., Sept. 25, 1868. d. Baltimore, Md., Nov. 2, 1889, age 57 years, 5 months. When Freedley wounded, Capt. Richard Gregory Lay commanded regiment. b. Washington, D.C., Feb. 6, 1834. Clerk in Washington, D.C. 2 Lt., 3rd U.S. Inf., June 30, 1859. M.O. Aug. 19, 1870. d. Washington, D.C., Sept. 21, 1919.

Number: 308   **Loss:** 6 k., 66 wd., 1 m.

Monument: Ayres Ave. Map reference **V C-9**

"July 2. Arrived in the morning and took position near the line of the Twelfth Corps. The regiment with the Brigade moved from the right to the left of the line and at 5 p.m. advanced across Plum Run near Little Round Top and supported the Second Brigade in its advance to the crest of the rocky wooded hill beyond and facing to the left engaged the Confederates but retired under a deadly fire on the left, right and rear after the Confederates had gained a position in the Wheatfield in the rear of the Brigade and took position on east slope of Little Round Top.

July 3. Remained in same position.

July 4. The Regiment with the Brigade made a reconnaissance and developed a force of the Confederate infantry and artillery in front."

4th UNITED STATES INFANTRY
4 Companies, CFHK
5th Corps, 2nd Div., 1st Brig.

Raised: Enlisted in New York City and Washington, D.C.

Organized: 1815. During April 1861, regimental headquarters was located at Fort Dalles on the Columbia River in Oregon State.

Commander: Capt. Julius Walker Adams, Jr. b. Westfield, Mass., April, 1840. Entered West Point from residence in Lexington, Kentucky. Grad. 1861. Lt., 4th U.S. Inf., June 24, 1861. Wd. June 27, 1862. M.O. June 25, 1864. d. Brooklyn, N.Y., Nov. 15, 1865.

Number: 179   **Loss:** 10 k., 30 wd.

Monument: Ayres Ave. Map reference **V C-9**

"July 2. Arrived in the morning and took position near the line of the Twelfth Corps. The Regiment with the Brigade moved from the right to the left of the line and at 5 p.m. advanced across Plum Run near Little Round Top and supported the Second Brigade in its advance to the crest of the rocky wooded hill beyond and facing to the left engaged the Confederates but retired under a deadly fire on both flanks and from the rear after the Confederates had gained a position in the Wheatfield in rear of the Brigade.

July 3. Remained in same position.

July 4. The Regiment with the Brigade made a reconnaisance and developed a force of the Confederate infantry and artillery in front and engaged on the skirmish line well to the front."

6th UNITED STATES INFANTRY
5 Companies, DFGHI
5th Corps, 2nd Div., 1st Brig.

Raised: Enlisted New York City and Suffolk County, Mass.

Organized: 1815. April 12, 1861 headquarters was at Benicia Barracks, California.

Commander: Capt. Levi Clark Bootes. b. Georgetown, D.C., Dec. 8, 1809. Entered U.S. Regular service as a private on July 19, 1846. Mexican War service. Transferred to 6th U.S. Inf. in 1848. Retd. with rank of Lt. Col., Oct. 7, 1874. d. Wilmington, Delaware, April 18, 1896.

Number: 236 **Loss:** 4 k., 40 wd.

Monument: Ayres Ave. Map reference **V D-8**

"July 2. Arrived in the morning and took position near the line of the Twelfth Corps. The Regiment with the Brigade moved from the right to the left of the line and at 5 p.m. advanced across Plum Run near Little Round Top and supported the Second Brigade in its advance to the crest of the rocky wooded hill beyond and facing to the left engaged the Confederates but retired under a deadly fire on both flanks and from the rear after the Confederates got possession of the Wheatfield in the rear of the Brigade and took position on Little Round Top.

July 3. Remained in same position.

July 4. The Regiment with the Brigade made a reconnaissance and developed a force of the Confederate infantry and artillery in front."

7th UNITED STATES INFANTRY
4 Companies, ABEI
5th Corps, 2nd Div., 2nd Brig.

Raised: Enlisted in New York City and Suffolk County, Mass.

Organized: 1815. April 12, 1861 headquarters was at Fort McLane, New Mexico territory.

Commander: Capt. David Porter Hancock. b. Wilkes Barre, Penn., Aug. 18, 1833. Entered West Point from home in Wilkes Barre on July 1, 1849. After Grad. he joined 7th U.S. as 2 Lt. d. in service with rank of Maj., in Harrisburg, Penn., May 21, 1880.

Number: 153 **Loss:** 12 k., 45 wd., 2 m.

Monument: Ayres Ave. Map reference **V C-8**

"July 2. Arrived in the morning and took position with the Brigade on the right of the Twelfth Corps. Later moved with the Brigade to the left and at 5 p.m. formed line on the right of Little Round Top, advanced across Plum Run and to the crest of the rocky wooded hill in front near the Wheatfield and facing to the left occupied the stone wall on the edge of the woods. The Confederates having opened fire on the right flank and advanced through the Wheatfield in the rear of the Brigade, was withdrawn under a deadly fire of musketry on both flanks and on the rear and of shot and shell from the Batteries, and formed in line on the right of Little Round Top.

July 3. Remained in same position.

July 4. Advanced nearly a mile in support of a skirmish line of the 12th and 14th Infantry."

8th UNITED STATES INFANTRY

At Taneytown Maryland during the battle.

10th UNITED STATES INFANTRY
3 Companies, DGH
5th Corps, 2nd Div., 2nd Brig.

Raised: Not enough information available.

Organized: Carlisle Barracks, Penn., 1855. April 12, 1861 headquarters at Ft. Laramie, Nebraska territory.

Commander: Capt. William Clinton. b. Philadelphia, Penn., 1817 or 1818. Mexican War service. Lt., 10th U.S. Inf., Mar. 3, 1855. M.O. Maj., Jan. 1, 1871.

Number: 106 **Loss:** 16 k., 32 wd., 3 m.

Monument: Ayres Ave. Map reference **V D-8**

"July 2. Arrived with the Brigade in the morning and took position on the right of the Twelfth Corps. Later moved to the left and at 5 p.m. the Brigade formed line with left on north slope of Little Round Top the right extending into the woods. Advanced across Plum Run and to the crest of the rocky wooded hill in front near the Wheatfield and facing left occupied the stone wall on the edge of the woods. The Confederates having opened fire on the right flank and advanced through the Wheatfield in the rear the Brigade was withdrawn under a heavy infantry fire on both flanks and from the rear and shot and shell from the Batteries, and was formed in line on the right of Little Round Top.

July 3. Remained in same position."

11th UNITED STATES INFANTRY
6 Companies, 1st Battalion, BCDEFG
5th Corps, 2nd Div., 2 Brig.

Raised: Enlisted in Suffolk County, Mass.; Marion County, Indiana; Des Moines County, Iowa; and St. Lawrence County, New York.

Organized: Formed at Ft. Independence, Boston Harbor, Mass., 1861.

Commander: Maj. Delancey Floyd-Jones. b. South Oyster Bay, Long Island, N.Y., Jan. 20, 1826. Grad. West Point, 1846. Mexican War service. Joined 11th U.S. as Maj., May 14, 1861. Retired as Col., 3rd U.S. Infantry, Mar. 20, 1879. d. Park Avenue Hotel, New York City, Jan. 19, 1902.

Number: 354 **Loss:** 19 k., 92 wd., 9 m.

Monument: Ayres Ave. Map reference **V D-8**

"July 2. Arrived in the morning with the Brigade and took position on the right of the Twelfth Corps. Afterwards moved to the left and at 5 p.m. formed line on the right of Little Round Top and advanced across Plum Run and to the crest of the rocky wooded hill in front under a fire of sharpshooters on the left and faced to the left with the Wheatfield on the right and rear. The Confederates having opened fire on the right flank and advancing through the Wheatfield in the rear, the regiment with the Brigade was withdrawn under a heavy fire of musketry and artillery and formed in line at the right of Little Round top.

July 3. Remained in same position."

12th UNITED STATES INFANTRY
8 Companies, 1st Battalion, ABCDG, 2nd Battalion, ACD
5th Corps, 2nd Div., 1st Brig.

Raised: Enlisted in New York City and counties of Dubuque, Iowa; Cass, Indiana; Essex, New Jersey; and Albany, New York.

Organized: Formed at Fort Hamilton, New York Harbor, N.Y., 1861.

Commander: Capt. Thomas Searle Dunn. b. Hanover, Indiana, 1822 or 1823. Attnd. Hanover College, Indiana. Mexican War service. Farmer in Logansport, Indiana. Capt., 9th Indiana Infantry (3 mos.). Capt., 12th U.S. Inf., May 14, 1861. Retd. as Maj., June 29, 1878. d. Santa Monica, California, Nov. 14, 1895.

Number: 453 **Loss:** 8 k., 71 wd., 13 m.

Monument: Ayres Ave. Map reference **V D-8**

"July 2. Arrived in the morning and took position with the Brigade and Division near the Twelfth Corps on the right. Moved with the Division from the right to the left of the line and at 5 p.m. with the Brigade moved across Plum Run near Little Round Top and supported the Second Brigade in its advance to the crest of the rocky wooded hill in front and facing left engaged the Confederates but retired under a heavy fire on both flanks and from the rear after the Confederates had obtained possession of the Wheatfield in the rear of the Brigade and went into position on Little Round Top.

July 3. Remained in same position.

July 4. Regiment with the 14th supported the 3rd, 4th, and 6th as infantry in a reconnaissance and developed a force of the Confederate infantry and artillery in front."

14th UNITED STATES INFANTRY
8 Companies, 1st Battalion, ABDEFG, 2nd Battalion, FG
5th Corps, 2nd Div., 1st Brig.

Raised: Enlisted in New York City. Also counties of Providence, R.I.; Onondaga, New York; and Chemung, New York.

Organized: Formed at Fort Trumbull, New London, Conn., 1861.

Commander: Maj. Grotius Reed Giddings. Father was prominent Whig U.S. Congressman, Joshua R. Giddings. Grotius was b. Jefferson, Ohio, June 21, 1834. Lawyer in Jefferson. Capt., 23rd Ohio Inf. Maj., 14th U.S. Inf., to date May 14, 1861. d. Lt. Col., of fever at Macon, Ga., June 21, 1867.

Number: 601 **Loss:** 18 k., 110 wd., 4 m.

Monument: Crawford Ave. Map reference **V D-8**

"July 2. Arrived in the morning and took position with the Brigade and Division near the Twelfth Corps on the right. Moved with the Division from the right to the left of the line at 5 p.m. with the Brigade moved across Plum Run near Little Round Top and supported the Second Brigade in its advance to the crest of the rocky wooded hill beyond and facing left engaged the Confederates but retired under a heavy fire on both flanks and from the rear after the Confederates had possession of the Wheatfield in the rear of the Brigade and went into position on Little Round Top. July 3. Remained in same position.

July 4. The regiment with the 12th supported the 3rd, 4th and 6th U.S. Infantry in a reconnaissance and developed a force of Confederate infantry and artillery in front."

17th UNITED STATES INFANTRY
7 Companies, 1st Battalions, ACDGH, 2nd Battalion, AB
5th Corps, 2nd Div., 2nd Brig.

Other Names: "Maine Regulars"

Raised: Enlisted in counties of Aroostook, Penobscot and Cumberland, Maine. Also Erie County, New York and Wayne County, Michigan.

Organized: Formed at Fort Preble, Cape Elizabeth, Maine, 1861.

Commander: Lt. Col. James Durell Greene. b. Lynn, Mass., May 12, 1828. Grad. Harvard University, 1849. Officer in Mass. Militia. Inventor and manufacturer of breech loading rifle in Cambridge, Mass. when the war began. Lt. Col., 5th Mass. Lt. Col., 17th U.S. Inf., May 14, 1861. Resigned with the rank of Col., June 25, 1867. d. Ypsilanti, Mich., Mar. 21, 1902.

Number: 334 **Loss:** 25 k., 118 wd., 7 m.

Monument: Ayres Ave. Map reference **V D-8**

"July 2. Arrived in the morning and took position with the Brigade on the right of the Twelfth Corps. Later moved to the left and at 5 p.m. formed line with the Brigade at the right of Little Round Top and advanced across Plum Run to the crest of the rocky wooded hill beyond near the Wheatfield under a severe fire from the Confederate sharpshooters on the left then facing left the Regiment with the Brigade occupied the stone wall on the edge of the woods. The Confederates, having opened fire on the right and advanced in the Wheatfield in the rear, the Brigade was withdrawn under a heavy fire on both flanks and from the rear and formed in line on the right of Little Round Top having been engaged about two hours.

July 3. Remained in same position."

1st UNITED STATES SHARPSHOOTERS
3rd Corps, 1st Div., 2nd Brig.

Raised: States of New York (ABDH), Michigan (CIK), New Hampshire (E), Vermont (F), Wisconsin (G).

Organized: Weehawken, N.J. M.I. Nov. 1861.

Commanders: Col. Hiram Berdan. b. Phelps, N.Y., Sept. 6, 1824. Inventor and manufacturer. 1861 living on Long Island, N.Y. Col., 1st U.S.S.S., Nov. 30, 1861. Wd. Aug. 30, 1862. M.O. Jan. 2, 1864. d. Washington, D.C., Mar. 31, 1893. Buried Arlington National Cemetery. When Berdan took command of brigade, Lt. Col. Casper Trepp led the regiment. b. Splugen, Canton Graubunden, Switzerland. Officer in Hamburg Legion during Crimea War.

151

Afterwards came to U.S. Architect in New York City. Capt., Co. A, 1st U.S.S.S., Sept. 22, 1862. Killed during Mine Run campaign, Va., Nov. 30, 1863.
Number: 371   **Loss:** 6 k., 37 wd., 6 m.
Monuments and Markers:
Companies CIK, Little Round Top. Map reference **V F-10**
Company E, Hancock Ave. Map reference **II E-17**
Companies ABDH, Berdan Ave. Map reference Map **IV** inset
Company F, Berdan Ave. Map reference Map **IV** inset
Company G plus marker. Emmitsburg Road. Map reference **IV D-5**

2nd UNITED STATES SHARPSHOOTERS
8 Companies, ABCDEHFG
3rd Corps, 1st Div., 2nd Brig.

Raised: Minnesota (A), Michigan (B), Pennsylvania (C), Maine (D), Vermont (EH), New Hampshire (FG).
Organized: Formed as a unit in Washington, D.C. M.I. Oct.-Dec., 1861.
Commander: Maj. Homer Richard Stoughton. b. Quechee, Vermont, Nov. 13, 1836. Railroad agent in Randolph, Vt. Capt., Co. E, 2nd U.S.S.S., Nov. 9, 1861. M.O. Lt. Col., Jan. 23, 1865. d. Sept. 17, 1902, at brother's home in Otsego, N.Y. Residence, Barre, Vt.
Number: 200   **Loss:** 5 k., 23 wd., 15 m.
Monuments and Markers:
Company D, Bushman House. Map reference **V G-2**
Company B, Little Round Top. Same location as Cos. CIK, 1st U.S.S.S. Map reference **V F-10**
Companies FG, Hancock Ave. Same As Co. E, 1st U.S.S.S.
    Map reference **II E-17**
Companies EH, near Slyder House. Map reference **V H-4**

1st UNITED STATES ARTILLERY,
BATTERIES E AND G CONSOLIDATED
Cavalry Corps, 2nd Div.

Raised: Enlisted in New York City and Suffolk County, Mass.
Organized: 1821. April 12, 1861 Battery E at Ft. Sumpter and Battery G was at Ft. Pickens, Florida. Consolidated in Feb. 1862.
Commander: Capt. Alanson Merwin Randol. b. Newburgh, N.Y., Oct. 23, 1837. Grad. West Point, 1860. 2 Lt., 4th U.S. Artillery. Assigned to 1st U.S. Artillery, Nov. 22, 1860. Col., 2nd New York Cavalry, Dec. 1864. M.O. Volunteer service, June 23, 1865. d. Maj. of 3rd U.S. Artillery, at New Almaden, California, May 7, 1887. Buried San Francisco National Cemetery, California.
Number: 4 Ordnance Rifles. 84 men.   **Loss:** None
Markers: East Cavalry Battlefield. No map reference.
"July 1 & 2. With First Brigade Second Cavalry Division. Not engaged.
July 3. One section under Lieut. James Chester was ordered to Second Bri-

gade, Third Cavalry Division and took position west of the Low Dutch Road and with Brig. General Custer's Second Brigade, Third Division Cavalry Corps was hotly engaged in repelling the attack of Major General Stuart's Confederate Cavalry Division. The one section under Lieut. Ernest L. Kinney remained near the Hanover Road."

1st UNITED STATES ARTILLERY, BATTERY H
Artillery Reserve, 1st Regular Brigade

Other Names: "Randol's"

Raised: Enlisted in New York City and counties of Suffolk, Mass. and Hamilton, Ohio.

Organized: 1821. On April 12,1861 was at Ft. Sumter with 4 officers and 32 men.

Commanders: Lt. Chandler Price Eakin. b. Philadelphia, Penn., Dec. 26, 1836. Civil Engineer in Philadelphia. Pvt., Independent Company, Penn. Artillery (3 mos.). 2 Lt., 1st U.S. Artillery, Aug. 5, 1861. Wd. May 5, 1862 and July 2, 1863. Retired as Capt., 1st U.S. Artillery, Jan. 14, 1888. d. Philadelphia, Nov. 13, 1903. When Eakin wd. at Gettysburg, 2 Lt. Philip D. Mason commanded battery. b. Boston, Massachusetts, Feb. 2, 1842. 2 Lt., 1st U.S. Artillery, Mar. 18, 1862. d. as 1st Lt., July 18, 1864 of wounds received June 11, 1864 at Battle of Trevilian Station, Va.

Number: 6 12-pdr. Napoleons. 101 men.   **Loss:** 1 k., 8 wd., 1 m.

Marker: National Cemetery. Map reference II I-5

"July 2. In position on Cemetery Hill facing the Emmitsburg Road. Engaged July 2nd and 3rd. Lieut. Eakin was severely wounded after his guns went into battery and the command devolved on Lieut. Philip D. Mason."

1st UNITED STATES ARTILLERY, BATTERY I
2nd Corps Artillery Brigade

Other Names: "Ricketts'"

Raised: Enlisted in New York City and Suffolk County, Mass.

Organized: 1821. April 12, 1861 was at Ft. Leavenworth, Kansas.

Commanders: Lt. George Augustus Woodruff. b. Marshall, Michigan, May 27, 1840. Grad. West Point, 1861. Assigned to 1st U.S. Artillery. d. July 4, 1863 of wounds received on July 3. General W. S. Hancock wrote, "Among all the brave men who fell at Gettysburg there are none whose loss I regret more than his." When Woodruff fell, Lt. Tully McCrea led the battery. b. Natchez, Mississippi, July 23, 1839. Lived in Urbana, Ohio before entering West Point. Grad. 1862. Assigned to 1st U.S. Artillery. Wd., Feb. 20, 1864. Retired as Brig. Gen., Feb. 22, 1903. d. West Point, New York, Sept. 5, 1918.

Number: 6 12-pdr. Napoleons. 113 men.   **Loss:** 1 k., 24 wd.

Monument: Hancock Ave. Map reference II F-8

"July 2 & 3. Arrived and took position in Ziegler's Grove on the left of Evergreen Cemetery actively engaged and assisted in repelling Longstreet's assault. Lieut. Woodruff was mortally wounded on the 3rd and the command devolved on Lieut. Tully McCrea."

1st UNITED STATES ARTILLERY, BATTERY K
Cavalry Corps, 1st Div.

Raised: Enlisted in cities of New York, Baltimore and Washington, D.C.
Organized: 1838. April 12, 1861 was at Ft. Taylor, Key West, Florida.
Commander: Capt. William Montrose Graham. Nephew on mother's side of General George G. Meade. b. Washington, D.C., Sept. 28, 1834. 1853 served as astronomer in the expedition of Gov. Isaac Stevens to the Pacific coast. 2 Lt., 1st U.S. Artillery, June 7, 1855. 1865 served as Col. of 2nd District of Columbia Infantry until M.O. Sept. 12, 1865. Retired from Regular service Sept. 28, 1898. d. Annapolis, Maryland, Jan. 16, 1916.
Number: 6 Ordnance Rifles. 122 men.   **Loss:** 2 k., 1 wd.
Monument: Emmitsburg Road. South Cavalry Battlefield. Map reference see Key map.
"July 3. Arrived on the field and took position on the left with cavalry and engaged during the attack of Brig. General E. J. Farnsworth's and Brig. General W. Merritt's Brigades on the Confederate right."

2nd UNITED STATES ARTILLERY, BATTERY A
Cavalry Corps, 1st Div.

Other Names: "Tidball's"
Raised: Enlisted in New York City and Hamilton County, Ohio.
Organized: 1821. April 4, 1861 left Washington, D.C. for New York City, then Ft. Pickens, Florida, arriving April 19.
Commander: Lt. John Haskell Calef. b. Gloucester, Mass., Sept. 24, 1841. Grad. West Point, 1862. Transferred to 2nd U.S. Artillery from 5th U.S. Artillery, Oct. 6,1862. Retd. as Lt. Col., Aug. 1900. d. St. Louis, Missouri, Jan. 14, 1912.
Number: 6 Ordnance Rifles. 80 men.   **Loss:** 12 wd.
Monument: Chambersburg Pike. Map reference I G-4
The first Union artillery shot fired in the battle came from Calef's gun #233, today mounted at the base of the the Buford monument on the Chambersburg Pike.
"July 30. Arrived in the evening from Emmitsburg and took position on the Chambersburg Pike.
July 1. Advanced with the Cavalry went into position with right section on right of the road left section on the left and centre section with Col. Wm. Gamble's Brigade on the right of the Fairfield Road. The First Union gun of the battle was fired from right section and the positions held under a severe fire until the First Corps arrived about 10 a.m. The Battery was then relieved by Capt. J. A. Hall's 2nd Maine Battery and after being supplied with ammunition returned about 3 p.m. But under a front and enfilading fire it retired

to a line in front of Cemetery Ridge and towards night moved to the left about a mile and bivouacked for the night near the Third Corps. July 2, a.m. Marched with the First Brigade of Major General John Buford's Division to Taneytown en route to Westminster."

Marker: Section marker on Reynolds Ave. Map reference I J-4

2nd UNITED STATES ARTILLERY, BATTERIES B AND L CONSOLIDATED
Cavalry Corps, Reserve Artillery Brigade

Other Names: L-"Hayes' " B-"Robertson's"

Raised: Enlisted in New York City and Hamilton County, Ohio.

Organized: B-1821. L-1847. Both batteries at Fort Monroe, Virginia on April 12, 1861.

Commander: Lt. Edward Heaton. b. Cincinnati, Oh., Sept. 29, 1842. Clerk in Cincinnati. 2 Lt., 2nd U.S. Artillery, Aug. 23, 1861. Joined battery May 1863. Resigned Mar. 1, 1865 to continue education. Grad. Yale College, 1869. Lawyer at death in Ridgefield, N.J., January 12, 1884.

Number: 6 Ordnance Rifles. 136 men.   **Loss:** None

Monuments: Granite School Road and Pleasonton Ave. Map references III D-19, II J-17

Pleasonton Ave—"July 2. Arrived near the battlefield at 5:30 a.m. and reported to Major General Alfred Pleasonton who ordered the Battery to be held in reserve until near dark when it was moved back two miles on the Baltimore Pike for the night.

July 3. Moved to the front and was ordered to the position occupied the day before but being subject to the severe artillery fire the Battery was ordered to retire out of range and there remained until the close of the battle."

2nd UNITED STATES ARTILLERY, BATTERY D
6th Corps Artillery Brigade

Other Names: "Platt's"

Raised: Enlisted in New York City and Hamilton County, Ohio.

Organized: 1821. April 1861 was at U.S. Arsenal at Fayetteville, North Carolina.

Commander: Lt. Edward Bancroft Williston. b. Norwich, Vermont, July 15, 1836. Attnd. Norwich University, Vt. Civil engineer living in San Francisco, California in 1861. 2 Lt., 2nd U.S. Artillery, Aug. 5, 1861. Awarded Medal of Honor for gallantry at Trevilian Station, Va., June 12, 1864. Retd., Col. in Regular service, July 15, 1900. d. San Diego, California, April 24, 1920. Buried Arlington National Cemetery.

Number: 6 12-pdr. Napoleons. 112 men.   **Loss:** None

Monument: Taneytown Road. Map reference, see key map.

"July 2. Arrived with the Corps and took position and remained on Taneytown Road. Not engaged.

2nd UNITED STATES ARTILLERY, BATTERY G
6th Corps Artillery Brigade

Other Names: "Thompson's"
Raised: Enlisted in New York City and Suffolk County, Mass.
Organized: 1821. April 12, 1861 located at Ft. Mackinac, Michigan.
Commander: Lt. John Hartwell Butler. b. Worcester, Mass., Oct. 15, 1838.
   Druggist in Hartford, Conn.2 Lt., 2nd U.S. Artillery, May 22, 1861. Wd. Nov.
   7, 1863 and Nov. 8, 1863. Left foot amputated. M.O. Feb. 7, 1865. Capt., 42nd
   U.S. Inf., Jan. 22,1867. Retd. May 14, 1870. d. Hartford, Conn., Jan. 23, 1910.
Number: 6 12-pdr. Napoleons. 113 men.   **Loss:** None
Monument: Near Cyclorama Center. Map reference **II G-8**
   "July 2. Arrived in the afternoon with the Corps and held in reserve.
   July 3. Brought up to Ziegler's Grove in rear of Third Division Second Corps
   on repulse of Longstreet's assault."

2nd UNITED STATES ARTILLERY, BATTERY M
Cavalry Corps, 3rd Division

Other Names: "Hartsuff's"
Raised: Enlisted in New York City and Hamilton County, Ohio.
Organized: 1847. April 12, 1861 was en route from N.Y. to Ft. Pickens, Florida.
Commander: Lt. Alexander Cummings McWhorter Pennington, Jr. b. Newark,
   N.J., Jan. 8, 1838. Grad. West Point, 1860. Assigned 2nd U.S. Artillery. Wd.
   Nov. 2, 1862. Col., 3rd N.J. Volunteer Cavalry, Oct. 1, 1864. M.O. Aug. 1,
   1865. Continued in Regular service until retirement as Brig. Gen., Oct. 17,
   1899. d. New York City, Nov. 30, 1917.
Number: 6 Ordnance Rifles. 126 men.   **Loss:** 1 wd.
Monument: East Cavalry Battlefield. No map reference.
   "July 2. Engaged with the Confederates at Hunterstown.
   July 3. Engaged in Brig. Gen. Custer's Brigade with Maj. Gen. J.E.B. Stuart's
   Confederate Cavalry on the right of the Union Army."

3rd UNITED STATES ARTILLERY, BATTERY C

Detached with Huey's Cavalry Brigade in Maryland during the Battle of
Gettysburg.

3rd UNITED STATES ARTILLERY,
BATTERIES F AND K CONSOLIDATED
Artillery Reserve, 1st Regular Brigade

Other Names: F-"Beckwith's" K-"Livingston's"
Raised: Enlisted, for the most part, in New York City.
Organized: F-1821, K-1838. April 12, 1861, both batteries at Fort Monroe, Virginia.

Commander: Lt. John Graham Turnbull. b. Washington, D.C., Sept. 9, 1843. Father was friend of General Winfield Scott. 2 Lt., 3rd U.S. Artillery, Aug. 5, 1861. Retired as Maj., Aug. 27, 1896. d. Washington, D.C., May 8, 1898.

Number: 6 12-pdr. Napoleons. 145 men. **Loss:** 9 k., 14 wd., 1 m.

Monuments: Near Meade's Headquarters and Emmitsburg Road. Map references **II F-10, IV D-5**

"July 1. Took position on crest of hill near General Meade's Headquarters. July 2. Moved to a position on the right of a log house on the Emmitsburg road with General A. A. Humphreys' Division Third Corps. Engaged here but was compelled to retire with the loss of 45 horses killed and 4 guns which were afterwards recaptured.

July 3. Went into position near the Taneytown Road on the left of Cemetery Hill."

4th UNITED STATES ARTILLERY, BATTERY A
2nd Corps Artillery Brigade

Other Names: "Morgan's"

Raised: At Gettysburg, very few members had enlisted in Regular service. The battery had been brought up to strength in Oct. 1862 with the addition of over 100 volunteers, predominantly from the 4th Ohio Infantry.

Organized: 1821. On April 12, 1861, battery was at Fort Crittenden, Utah Territory.

Commanders: Lt. Alonzo Hereford Cushing. b. near Delafield, Wisconsin, Jan. 19, 1841. Worked in grocery store in Fredonia, N.Y. Grad. West Point, 1861. Assigned to 4th U.S. Artillery. Killed July 3, 1863 at Gettysburg. When Cushing killed, Sergt. William Frederick Fuger assumed command of battery. b. Goppingen, Germany, June 18, 1836. Attnd. University of Tubingen. At age 18, "inheriting but little of the ancient family plunder, but something of their ambition and persistence," he set out for America. Clerk in New York City. Pvt., 4th U.S. Artillery, Aug. 21, 1856. Wd. June 30, 1862 and Sept. 17, 1862. Awarded Medal of Honor for action at Gettysburg. Retired as Maj., June 18, 1900. d. Washington, D.C., Oct. 13, 1913. Buried Arlington National Cemetery.

Number: 6 Ordnance Rifles. 129 men. **Loss:** 6 k., 32 wd.

Monument: Hancock Ave. Map reference **II E-12**

Marker: In honor of Cushing. Located near guns. Fuger's account of the fight on July 3, 1863:

"Of this bombardment I can only say it was the most terrific cannonade I ever witnessed, in fact, the most terrible the new world has ever seen, and the most prolonged. The earth shook beneath our very feet, and the hills and woods seemed to reel like a drunken man. For one hour and a half this terrific firing was continued, during which time the shrieking of shells, the fragments of rocks flying through the air, shattered from stone fences in front of Battery A, the splash of bursting shells and shrapnel, and the fierce neighing of wounded artillery horses, made a picture terribly grand and sublime. About 2:30 p.m. the order 'cease firing' was given; this was followed by cessation of the enemy's fire.

In this engagement all our ammunition was expended excepting canister. I will have to say that when our artillery ceased firing Gen. Alex. S. Webb came up to where Cushing was standing and said to him, 'Cushing, it is my opinion that the Confederate infantry will now advance and attack our position.' Cushing then said, 'I had better run my guns right up to the stone fence and bring all my canister alongside of each piece.' General Webb replied, 'All right, do so.'

Lieutenant Cushing ordered me to run the guns by hand to the stone fence, which was done at once, leaving room enough for nos. 1 and 2 to work. All the canister was piled up in rear of no. 2; in doing this we were obliged to take a closer interval, say about nine yards, owing to some obstructions toward our left, and on our right were two stone fences at right angles with each other (now called 'the Bloody Angle').

When the enemy was within about 400 yards Battery A opened with single charges of canister. At that time Cushing was wounded in the right shoulder, and within a few seconds after that he was wounded in the testicles - a very severe and painful wound.

He called me and told me to stand by him so that I could impart his orders to the battery; he became very ill and suffered frightfully. I wanted him to go to the rear. 'No,' he said, 'I stay right here and fight it out or die in the attempt.' When the enemy got within 200 yards double and treble charges of canister were used; those charges opened immense gaps in the Confederate lines. Lieutenant Milne, who commanded the right half battery, was killed when the enemy was within 200 yards of the battery. When the enemy came within about 100 yards from the battery Lieutenant Cushing was shot through the mouth and instantly killed. When I saw him fall forward I caught him with my arms, ordered two men to take his body to the rear; that placed me in command of the battery and I shouted to my men to obey my command and fire treble charges of canister; but still the Confederates came on. Owing to the dense smoke I could not see very far to the front, but to my utter astonishment I saw General Armistead leap over the stone fence with quite a number of his men (landing right in the midst of our battery), but my devoted cannoneers and drivers stood their ground, fighting hand-to-hand with pistols, sabers, handspikes and rammers, and with the arrival of the Philadelphia Brigade, commanded by the gallant General Webb, the enemy collapsed and Pickett's charge was defeated. I want to say here that we had about twenty-eight handspikes, which were all brought up to the guns; those handspikes are the finest weapons for close contact.

General Armistead fell, mortally wounded, where I stood, about seven yards from where Lieutenant Cushing, his young and gallant adversary, was killed."

4th UNITED STATES ARTILLERY, BATTERY B
1st Corps Artillery Brigade

Other Names: "Gibbon's"

Raised: At Gettysburg only 11 men of battery had originally joined as U.S. Regulars. The rest joined from volunteer ranks in the fall of 1862, coming from the 7th, 6th, and 2nd Wisconsin Inf., 24th Michigan Inf., and 19th Indiana Inf.

Organized: 1821. April 12, 1861 the battery was stationed at Fort Crittenden, Utah Territory.

Commander: Lt. James Stewart. b. Edinburgh, Scotland, May 18, 1826. Came to U.S. in 1844. Printer. Pvt., 4th U.S. Artillery, Oct. 29, 1851. Achieved rank of Lt. to date July 3, 1863. Wd. July 2, 1863. Retd. with rank of Capt., Mar. 20, 1879. d. Fort Thomas, Kentucky, April 19, 1905. Buried Arlington National Cemetery.

Number: 6 12-pdr. Napoleons. 90 men. **Loss:** 2 k., 31 wd., 3 m.

Monument: East Cemetery Hill. Map reference II K-3

"July 1. In position about 200 yards south of the Seminary until 3 p.m. when ordered to the support of Brig. General J. C. Robinson's Division First Corps and took position on Seminary Ridge one half the Battery between the Chambersburg Pike and the Railroad cut. The other half north of the cut in the corner of the woods was actively engaged. The Battery afterwards retired with the troops to Cemetery Hill where it went into position on the Baltimore Pike opposite the Evergreen Cemetery commanding the approach from the town. Two guns on the pike and two in the field. Two guns having been disabled.

July 2 & 3. Remained in this position.

Marker: Battery section along Chambersburg Pike. Map reference I H-6

4th UNITED STATES ARTILLERY, BATTERY C
Artillery Reserve, 1st Regular Brigade

Other Names: "Beach's"

Raised: New York City. In the fall of 1862 the battery took to the field with the addition of over 100 volunteers, many from the 14th Indiana Infantry.

Organized: 1821. April 12, 1861 stationed at Fort Crittenden, Utah Territory.

Commander: Lt. Evan Thomas. Son of Lorenzo Thomas, the Adjutant General of the United States Army. Evan b. Georgetown, D.C., May 29, 1843. 2 Lt., 4th U.S. Art., April 9, 1861. Killed with the rank of Capt. while fighting the Modoc Indians in California, April 26, 1873. Just before the end, finding his command trapped with no escape, Evan called out to his panic stricken men, "It is as good a place to die as any - fight and die like men and soldiers." Buried Arlington National Cemetery.

Number: 6 12-pdr. Napoleons. 112 men.   **Loss:** 1 k., 17 wd.

Monument: Hancock Ave. Map reference II E-17

"July 2. Arrived and took position on crest of hill near General Meade's Headquarters on the left of the Second Corps and was actively engaged in repelling the attack of the Confederates.

July 3. In position near the left of the Second Corps line."

4th UNITED STATES ARTILLERY, BATTERY E
Cavalry Corps, 3rd Division

Other Names: "Clark's"
Raised: New York City and St. Louis County, Missouri.
Organized: 1821. April 12, 1861 at Fort Randall, Nebraska Territory.
Commander: Lt. Samuel Sherer Elder. b. Harrisburg, Penn., Sept. 10, 1830. Teacher before enlistment. Pvt., 2nd U.S. Artillery, June 1, 1853. d. with rank of Maj. at Fort Monroe, April 6, 1885.
Number: 4 Ordnance Rifles. 64 men.   **Loss:** 1 k.
Monument: Near Big Round Top. Map reference **V K-5**

"July 3. Arrived on the field and took position on a hill southwest of Round Top and engaged under Brig. General E. J. Farnsworth in the afternoon against the Confederate right."

4th UNITED STATES ARTILLERY, BATTERY F
12th Corps Artillery Brigade

Other Names: "Best's"
Raised: Cities of New York and Baltimore, Maryland.
Organized: 1821. April 12, 1861 at Fort Ridgely, Minnesota. Unit dates its history back to Alexander Hamilton's Provincial Company of Artillery during the Revolutionary War.
Commander: 2 Lt. Sylvanus Tunning Rugg. b. Taunton, Mass., 1834. Regular service, 1850-1854. Laborer and clerk in St. Paul, Minn. and Taunton, Mass. Pvt., 2nd U.S. Artillery, Dec. 22, 1857. Dismissed from service July 22, 1864. d. Minneapolis, Minn., May 4, 1881.
Number: 6 12-pdr. Napoleons. 120 men.   **Loss:** 1 wd.
Monument: Baltimore Pike. Map reference **III B-10**

"July 1. Approached Gettysburg on the Baltimore Pike to Two Taverns and took position to counteract any movements of the Confederates from toward Hanover. At noon moved to the Hanover Road and marched to within one and one half miles of Gettysburg.

July 2. Took position so as to command a gap between the First and Second Corps.

July 3. In position near the left of the Second Corps line."

4th UNITED STATES ARTILLERY, BATTERY G
11th Corps Artillery Brigade

Other Names: "Howe's"

Raised: Enlisted in counties of Cook, Ill., and St. Louis, Missouri.

Organized: 1821. April 12, 1861 stationed at Ft. Randall, Nebraska Territory.

Commanders: Lt. Bayard Wilkeson. One of the youngest officers in command at Gettysburg. Father Samuel Wilkeson was a staff writer with the New York Tribune and its war correspondent with the Army of the Potomac. Bayard b. Albany, N.Y., May 17, 1844. 2 Lt., 4th U.S. Artillery, Aug. 14. 1862. During the fight on July 1 at Gettysburg, Wilkeson sat on his horse forward of his guns to keep his cannoneers to their work. A solid shot from the enemy's artillery cut his leg off just below the knee. He made a tourniquet of his handkerchief and had himself carried between two of his guns and from there continued to direct their fire. He was carried to the Adams County poor house nearby in a blanket by four of his men. He died there that night. "Just before he expired, it is said, he asked for water; a canteen was brought to him; as he took it a wounded soldier lying next to him begged, 'For God's sake give me some!' He passed the canteen untouched to the man, who drank every drop it contained. Wilkeson smiled on the man, turned slightly, and expired."

Samuel Wilkeson, at Meade's Headquarters during the fight, claimed his son's body.

Bayard was posthumously breveted up through the ranks to Lt. Col. to date July 1, 1863. After Wilkeson, Lt. Eugene Adolphus Bancroft commanded battery. b. Boston, June 17, 1825. Bookkeeper in Chicago. Pvt., Sturgis Rifles, Illinois Volunteers. 2 Lt., 4th U.S. Artillery, Oct. 24, 1861. Retired with rank of Capt., June 17, 1889. d. New London, Conn., Dec. 18, 1910.

Number: 6 12-pdr. Napoleons. 122 men.   **Loss:** 2 k., 11 wd., 4 m.

Monuments: Howard Ave. and the National Cemetery. Map references **I B-15, II J-4**

Howard Ave.—"July 1. Arrived at Gettysburg about 11 a.m. Advanced and took position two sections on Barlow's Knoll the left section detached near Alms house. Engaged Confederate infantry and artillery on right and left. Lieut. Wilkeson fell early mortally wounded and the command devolved on Lieut. Bancroft.

The sections were compelled to change positions several times. Retired about 4 p.m. one section relieving a section of Battery I, 1st Ohio on Baltimore Street in covering the retreat. About 5 p.m. took position on Cemetery Hill.

July 2. Moved to rear of cemetery facing Baltimore Pike. In action at the cemetery from 4:30 a.m. until 7 p.m.

July 3. About 2 p.m. two sections were engaged in the cemetery until the repulse of the Confederates."

4th UNITED STATES ARTILLERY, BATTERY K
3rd Corps Artillery Brigade

Other Names: "De Russy's"

Raised: Enlisted in New York City and Suffolk County, Mass. At Gettysburg

60 volunteer soldiers, many from the 120th N.Y. Infantry, were attached to bring the battery up to strength.

Organized: 1838. April 12, 1861 stationed at Ft. Ridgely, Minnesota.

Commanders: Lt. Francis Webb Seeley. b. Ashtabula, Ohio, April 12, 1837. Carpenter in Wabasha County, Minnesota. Pvt., 3rd U.S. Artillery, Feb.15, 1855. 2 Lt., 4th U.S. Artillery, Feb. 4, 1861. Wd. July 2, 1863. Resigned with rank of Capt., Aug. 31, 1864. Post Master of Lake City, Minnesota. d. Sawtelle, California, Dec. 30, 1910. Buried Los Angeles National Cemetery. When Seeley wd., 2 Lt. Robert James took command. b. Down, Ireland, 1836 or 1837. Pvt., 3rd U.S. Artillery, 1856. Joined 4th U.S. Artillery, Jan. 10, 1861. Cashiered Aug. 19, 1865.

Number: 6 12-pdr. Napoleons. 134 men.

Monument: Emmitsburg Road. Map reference IV G-3

"July 1. Arrived at night and encamped in a field south of the town between the Emmitsburg and Taneytown Roads.

July 2. Went into position at 4 p.m. on the right of Smith's log house on Emmitsburg Road with Brig. General A. A. Humphreys' Division Third Corps and soon took position on the left of the log house and at the left of an apple orchard and opened fire on the Confederate infantry as it began to advance. Hotly engaged with the Confederate infantry and artillery in front and on the left until about 7 p.m. when forced to retire and took position on the line from the Evergreen Cemetery to Little Round Top. Lieut. Seeley having been wounded the command devolved on Lieut. Robert James.

July 3. Remained in the position of the previous night."

5th UNITED STATES ARTILLERY, BATTERY C
Artillery Reserve, 1st Regular Brigade

Other Names: "Hascall's"

Raised: Northampton County, Pennsylvania.

Organized: Camp Greble, near Harrisburg, Sept., 1861.

Commander: Lt. Gulian Verplanck Weir. b. West Point, N.Y., Dec. 28, 1837. Father Robert was professor of drawing at West Point. Pvt., 7th N.Y. State Militia, April 26, 1861. 2 Lt., 5th U.S. Artillery, May 14, 1861. d. with rank of Capt. at Ft. Hamilton, N.Y. from a self inflicted gunshot wound, July 18, 1886.

Number: 6 12-pdr. Napoleons. 123 men.   **Loss:** 2 k., 14 wd.

Monument: Hancock Ave. Map reference II F-15

"July 2. Arrived at Gettysburg from near Taneytown and in the afternoon was ordered to the front and by direction of Major General W. S. Hancock took position 500 yards farther to the front and by order of Brig. General John Gibbon opened fire on the Confederates on the left front. The Confederates in front advanced to within a few yards no infantry opposing. Three of the guns were captured by the Confederates and drawn off to the Emmitsburg Road but were recaptured by the 13th Vermont and another regiment.

July 3. In the rear of the line until Longstreet's assault was made when the Battery was moved up to Brig. General A. S. Webb's line and opened with canister at short range on the advancing Confederates. At 6:30 p.m. returned to the Artillery Reserve."

5th UNITED STATES ARTILLERY, BATTERY D
5th Corps Artillery Brigade

Other Names: "Griffin's." "West Point Battery."

Raised: Enlisted New York City and Suffolk County, Mass.

Organized: Jan. 7, 1861, Lt. Charles Griffin was given permission to form an artillery battery at West Point. 70 men were transferred from Regular artillery and dragoon units. Jan. 31, 1861, the battery left West Point for Washington, D.C. where it was attached to the 5th U.S. Artillery as Battery D on July 4, 1861.

Commanders: Lt. Charles Edward Hazlett. b. either Zanesville or Newark, Ohio, Oct. 1838. Entering West Point, July 1, 1855, he listed residence as Zanesville. 2 Lt., 2nd U.S. Cavalry, May 6, 1861. Lt., 5th U.S. Artillery, May 14, 1861. Killed July 2, 1863. Lt. Benjamin Franklin Rittenhouse took command from Hazlett. b. Berwick, Penn., Dec. 15, 1839. Clerk in Washington, D.C. 2 Lt., 5th U.S. Artillery, May 14, 1861. Wd. June 19, 1864. Retired with rank of Capt., Oct. 7, 1874. d. Philadelphia, Penn., Mar. 6, 1915. Buried Arlington National Cemetery.

Number: 6 10-pdr. Parrotts. 124 men.   **Loss:** 7 k., 6 wd.

Monument: Little Round Top. Map reference **V F-11**

Marker: Location where Hazlett fell. **V F-11**

Account by Rittenhouse:

"July 2. p.m. Hazlett's battery ordered up the east side of Little Round Top to the crest. "In less time than it takes to tell it, four guns were on the crest, where a rider would hardly dare go to-day; a few minutes later Hazlett got the fifth piece into position over huge rocks, and a little later he got the sixth piece fairly lifted into position by the cannoneers and the infantry. As each piece was unlimbered, it spoke for itself, for the country on the left and front was full of rebels, with their battle flags flying, and coming so rapidly, that it seemed almost impossible to stop them. They were making for Little Round Top, but we were there, and as puff after puff of smoke flew out from those six Parrott guns, our boys for a mile down to the right, though they could not hear them in the roar of battle, could see that we held that point, and that the stars and stripes were there to stay. When the smoke cleared away, and men had time to look around and inquire how matters stood with this or that command, it was found that death had not stayed his hand. Colonel O'Rorke of the One Hundred and Fortieth New York was no more, the gallant [General] Weed was dying, and the noble Hazlett, while kneeling beside him to receive his last words, was mortally wounded."

5th UNITED STATES ARTILLERY, BATTERY F
6th Corps Artillery Brigade

Other Names: "Ayres'"
Raised: Enlisted in New York City, Philadelphia, Penn. and Erie County, N.Y.
Organized: 1861 at Camp Barry near Washington, D.C.
Commander: Lt. Leonard Martin. b. Green Bay, Wisconsin, Aug. 26, 1838. In seminary school in Poughkeepsie, N.Y. when appointed to West Point in 1856. Grad. 1861. 2 Lt., 4th U.S. Artillery, May 6, 1861. Lt., 5th U.S. Artillery, May 14, 1861. Resigned April 3, 1866. Also served as Col., 51st Wisconsin Infantry for 4 months in 1865. d. Winnebago, Wisc., April 14, 1890.
Number: 6 10-pdr. Parrotts. 130 men.   **Loss:** None.
Monument: Hancock Ave. Map reference **II G-8**

"July 2. Arrived in the afternoon with the Corps and held in reserve.
July 3. Brought up to Ziegler's Grove in rear of Third Division Second Corps on the repulse of Longstreet's assault."

5th UNITED STATES ARTILLERY, BATTERY I
5th Corps Artillery Brigade

Other Names: "Weed's"
Raised: Enlisted New York City and Luzerne County, Penn.
Organized: Camp Greble near Harrisburg, Penn. M.I. Oct. 1861.
Commanders: Lt. Malbone Francis Watson. b. Catskill, N.Y., June 2, 1839. Grad. West Point, 1861. 2 Lt., 1st U.S. Cavalry. Lt., 5th U.S. Artillery, May 14, 1861. Wd. July 2, 1863. Right leg amputated. Retired as Capt., Sept. 18, 1868. d. Dayton, Ohio, Dec. 9, 1891. When Watson wounded, Lt. Charles Curtis MacConnell led the battery. b. Pittsburgh, Penn., July 4, 1840. Law clerk in Pittsburgh. 2 Lt., 5th U.S. Artillery, May 14, 1861. Retired with rank of Capt., April 18, 1883. d. Narragansett, R.I., Aug. 4,1908. Buried Arlington National Cemetery.
Number: 4 Ordnance Rifles. 78 men.   **Loss:** 1 k., 19 wd., 2 m.
Monument: United States Ave. Map reference **IV J-10**

"July 2. About 4:30 p.m. arrived and took position north of Little Round Top. 5:30 moved to the front of the Peach Orchard.
On the advance of the Confederates driving back the Infantry, the Battery was retired across Plum Run near the Trostle House and fired shell and canister at the approaching Confederates until the Battery, disabled by the loss of men and horses, was captured by the 21st Mississippi Infantry. It was almost immediately recaptured with the assistance of the 39th New York Infantry and, being unserviceable, was taken to the Artillery Brigade."

5th UNITED STATES ARTILLERY, BATTERY K
12th Corps Artillery Brigade

Other Names: "Bainbridge's"
Raised: Enlisted in Pennsylvania counties of Berks, Blair and Schuykill.

Organized: Camp Greble, Harrisburg, Penn. M.I. Sept., 1861.

Commander: Lt. David Hunter Kinzie. b. Chicago, Ill., Jan. 23, 1841. When entering West Point in 1859, his occupation was a student and his residence was Kansas Territory. Grad., 1861. 2 Lt., 5th U.S. Artillery, May 14, 1861. d. with rank of Col., Atlanta, Ga., July 5, 1904. Buried National Cemetery in Marietta, Ga.

Number: 4 12-pdr. Napoleons. 86 men. **Loss:** 5 wd.

Monuments: Baltimore Pike and summit of Culp's Hill. Map references **III B-11, III E-3**

"July 1. Marched to within a mile and half of Gettysburg.

July 2. At daylight took position to command a gap between the First and Twelfth Corps. At 5 p.m. one section was placed on the summit of Culp's Hill and assisted in silencing Confederate Batteries on Benner's Hill. At 6 p.m. rejoined the Battery at the foot of Powers Hill.

July 3. At 1 a.m. posted with Battery F, 4th U.S. Artillery on the south side of Baltimore Pike opposite the centre of the line of the Twelfth Corps. At 4:30 a.m. opened fire on the Confederates in possession of the line vacated by the Twelfth Corps the preceding night. Firing continued at intervals until after 10 a.m. when the Confederates were driven out. Remained in the same position exposed to the severe shelling which came over Cemetery Hill in the afternoon."

1st UNITED STATES CAVALRY
8 Companies, ABCEHILM
Cavalry Corps, 1st Div., Reserve Brig.

Raised: Enlisted in New York City and Philadelphia, Penn. Also counties of Hamilton and Franklin, Ohio, and Ohio County, West Virginia.

Organized: Originally a Dragoon unit formed in 1833 in response to the Black Hawk war. 1861 became 1st U.S. Cavalry. April 12, 1861 headquarters at Fort Tejon, California.

Commander: Capt. Richard S. C. Lord. b. Bellefontaine, Ohio, Oct. 26, 1832. Grad. West Point, 1856. 2 Lt., 7th U.S. Inf. Lt., 1st U.S. Cav., April 23, 1861. Wd. July 9, 1863. d. Bellefontaine, Oct. 15, 1866.

Number: 392 **Loss:** 1 k., 9 wd., 5 m.

Monument: 3 Stone markers, South Cavalry Battlefield. See key map.

"July 3. Moved with the Brigade at 12 m. under Brig. General W. Merritt from Emmitsburg and attacked the Confederate right and rear and was engaged for four hours until the action was brought to a close by a heavy rain."

2nd UNITED STATES CAVALRY
Cavalry Corps, 1st Div., Reserve Brig.

Raised: Enlisted Washington, D.C. and New York City. Many other areas represented.

Organized: 1836 as a Dragoon unit. Aug. 1861 became 2nd U.S. Cav. April 12, 1861 headquarters at Fort Crittenden, Utah Territory.

Commander: Capt. Theophilus Francis Rodenbough. b. Easton, Penn., Nov. 5, 1838. Attnd. Lafayette College, Penn. Clerk in Easton. 2 Lt., 2nd Dragoons, Mar. 27, 1861. Wd. June 11, 1864 and Sept. 19, 1864. Lost right arm. Also was Col., 18th Penn. Cav. for 7 months in 1865. Retired with rank of Col. Dec. 15, 1870. Awarded Medal of Honor for gallantry at Trevilion Station, Va. June 11, 1865. d. New York City, Dec. 19, 1912.

Number: 510   **Loss:** 3 k., 7 wd., 7 m.

Monument: 3 stone markers, South Cavalry Battlefield. See key map.

"July 3. Moved with the Brigade at 12 m. under Brig. General W. Merritt from Emmitsburg and attacked the Confederate right and rear and was engaged for four hours until the action was brought to a close by a heavy rain."

5th UNITED STATES CAVALRY
11 Companies (No Company L)
Cavalry Corps, 1st Div., Reserve Brig.

Raised: Cities of New York, Washington, D.C. and Baltimore, Md. Also Suffolk County, Mass. and Hamilton County, Ohio.

Organized: 1855 in Louisville, Kentucky as 2nd U.S. Cavalry. Became 5th U.S. Cav. Aug. 3, 1861. Headquarters left Ft. Mason, Texas, April 12, 1861 for Carlisle Barracks, Penn.

Commander: Capt. Julius Wilmot Mason. b. Penn., Jan. 19, 1835. Father from Towanda, Penn. Grad. Western Military Institute in Kentucky. Engineer and militia officer in Towanda. 2 Lt., 2nd U.S. Cav. Lt., 5th U.S. Cav., Aug. 3, 1861. d. on duty with rank of Maj., at Fort Huachuca, Arizona Territory, Dec. 19, 1882.

Number: 454   **Loss:** 4 wd., 1 m.

Monument: South Cavalry Battlefield. See key map.

"July 3. Moved with the Brigade at 12 m. under Brig. General W. Merritt from Emmitsburg and attacked the Confederate right and rear and was engaged for four hours until the action was brought to a close by a heavy rain."

6th UNITED STATES CAVALRY

Not at Gettysburg battlefield.

2nd VERMONT INFANTRY
6th Corps, 2nd Div., 2nd Brig.

Raised: Counties of Washington, Bennington, Rutland, Windham, Orange, Chittenden, Franklin, Windsor and Addison.

Organized: Camp Underwood, Burlington, Vt. M.I. June 20, 1861.

Commander: Col. James H. Walbridge. b. Bennington, Vt., July 29, 1826. "in his early youth he became a sailor before the mast. Later he was a citizen of California, a pioneer printer, a gold prospector, and a member of the famous vigilance committee that restored law and order in San Francisco in 1856." Farmer in Bennington at the start of the war. Capt., Co. A, 2nd Vt., May 14, 1861. M.O. April 1, 1864. d. North Bennington, Dec. 15, 1913.

Number: 528 **Loss:** None

Monument: No unit monument. Regiment honored on brigade monument, Wright Ave. Map reference **V J-14**

3rd VERMONT INFANTRY
6th Corps, 2nd Div., 2nd Brig.

Raised: Counties of Windsor, Caledonia, Essex and Orleans.

Organized: Camp Baxter, St. Johnsbury, Vt. M.I. July 16, 1861.

Commander: Col. Thomas Orville Seaver. b. Cavendish, Vt., Dec. 23, 1833. Attnd. Norwich University, Vt. Grad. Union College, N.Y., 1859. Law student in Pomfret, Vt. Capt., Co. F, 3rd Vt., July 16, 1861. M.O. July 27, 1864. Awarded Medal of Honor for gallantry near Spotsylvania Court House, Va., May 10, 1864. Judge in Woodstock, Vt. at time of death, July 11, 1912.

Number: 428 **Loss:** None

Monument: No unit monument. Regiment honored on brigade monument, Wright Ave. Map reference **V J-14**

4th VERMONT INFANTRY
6th Corps, 2nd Div., 2nd Brig.

Raised: Counties of Windsor, Bennington, Windham, Orange, Washington and Caledonia.

Organized: Camp Holbrook, Brattleboro, Vt. M.I. Sept. 21, 1861.

Commander: Col. Charles Bradley Stoughton. b. Chester, Vt., Oct. 31, 1841. Grad. Norwich University, Vt., 1861. Residence in Bellows Falls, Vermont. Lt. and Adjt., 4th Vt., Sept. 21, 1861. Lost right eye at Battle of Funkstown, Md., July 10, 1863. M.O. Feb. 2, 1864. Practiced law. d. old soldiers home in Bennington, Vt., Jan. 17, 1898.

Number: 437 **Loss:** 1 wd.

Monument: No unit monument. Regiment honored on brigade monument, Wright Ave. Map reference **V J-14**

5th VERMONT INFANTRY
6th Corps, 2nd Div., 2nd Brig.

Raised: Counties of Franklin, Addison, Rutland, and Chittenden.
Organized: Camp Holbrook, St. Albans, Vt. M.I. Sept. 16 and 17, 1861.
Commander: Lt. Col. John Randolph Lewis. b. Edinboro, Penn., Sept. 16, 1834. Attnd. Penn. College of Dental Surgery and University of Vermont Medical School. Dentist in Burlington, Vt. Sergt., 1st Vt. Inf. Capt., Co. I, 5th Vt., Sept. 16, 1861. Wd. June 30, 1862 and in the Battle of the Wilderness, Va. resulting in the amputation of his left arm. M.O. Col., Sept. 11, 1864. Served in Veteran Reserve Corps and as Maj. of 44th U.S. Infantry until retirement as Col., April 28, 1870. d. Chicago, Ill., Feb. 8, 1900. Buried Arlington National Cemetery.
Number: 341 **Loss:** None.
Monument: No unit monument. Regiment honored on brigade monument, Wright Ave. Map reference **V J-14**

6th VERMONT INFANTRY
6th Corps, 2nd Div., 2nd Brig.

Raised: Counties of Washington, Windsor, Addison, Chittenden and Franklin.
Organized: Camp Smith, Montpelier, Vt. M.I. Oct. 15, 1861.
Commander: Col. Elisha L. Barney. b. Swanton, Vt., April 13, 1832. Merchant in Swanton. Capt., Co. K, 6th Vt., Oct. 15, 1861. Wd. Sept. 14, 1862 and May 5, 1864. d. May 10, 1864 of wounds received May 5, 1864 in the Wilderness, Va.
Number: 362 **Loss:** None
Monument: No unit monument. Regiment honored on brigade monument, Wright Ave. Map reference **V J-14**

12th VERMONT INFANTRY

Guarding Trains at Rock Creek Church, Penn.

13th VERMONT INFANTRY
1st Corps, 3rd Div., 3rd Brig.

Raised: Counties of Washington, Chittenden, Lamoille and Franklin.
Organized: 9 month regiment formed in Brattleboro, Vt. M.I. Oct. 3,1862.
Commanders: Col. Francis Voltaire Randall. b. Braintree, Vt., Feb. 13, 1824. Member of state legislature. Lawyer in Montpelier at the start of the war. Capt., 2nd Vt. Inf. Col., 13th Vt., Oct. 10, 1862. M.O. July 21, 1863. d. Northfield Center, Mar. 1, 1885. When Randall elevated to brigade command, Lt. Col. William Day Munson led the regiment. b. Colchester, Vt., Feb. 17, 1833.

Grad. Norwich University, Vt., 1854. Farmer in Colchester. Capt., Co. D, 13th Vt., Sept. 6, 1862. Wd. July 3, 1863. M.O. July 21, 1863. d. Colchester, Oct. 28, 1903. When Munson wd., Maj. Joseph J. Boynton took command. b. Stowe, Vt., June 9, 1833. Farmer in Stowe. Capt., Co. E, 13th Vt., Oct. 10, 1862. M.O. July 21, 1863. After the war became a doctor. d. South Framingham, Mass., June 17, 1897.

Number: On June 27, the regiment had 710 men. **Loss:** 10 k., 103 wd., 10 m.

Monument: Hancock Ave. Figure represents Lt. Stephen Flavius Brown (1841-1903. Farmer from Swanton, Vt.). On the march to Gettysburg, he ordered a guard protecting a well to stand aside so some of his men could quench their thirst. He was placed under arrest and relieved of his sword. When his unit went into action, he was allowed to go with them, carrying a common camp hatchet. During Pickett's charge he obtained a sword from a captured Confederate officer. After the battle, charges against him were dropped and he continued in service until left arm amputated after being shot in the Battle of the Wilderness, May 6, 1864. When the veterans submitted their design for their monument, Lt. Brown was shown carrying the hatchet. The battlefield commissioners protested and a compromise was reached where the figure would carry the sword in a scabbard, and have the camp hatchet at his feet. The sword is an accurate representation of the one captured by Brown which he kept after the war as a souvenir.

"July 2. Five companies under Lieut.-Colonel Wm. D. Munson supported Batteries on Cemetery Hill. Near evening the other five companies commanded by Colonel Francis V. Randall charged to the Rogers House on the Emmitsburg Road, captured 83 prisoners and recapturing 4 guns after which they took position here and were soon joined by the five companies from Cemetery Hill.

July 3. In the morning 100 men advanced 45 yards under the fire of sharp-shooters and placed a line of rails. When the Confederate column crossed the Emmitsburg Road the regiment advanced to the rail breastworks and opened fire as the Confederates obliqued to their left. The regiment changed front forward on first company advanced 200 yards attacking the Confederate right flank throwing it into confusion and capturing 243 prisoners." Map reference **II E-15**

Markers: Markers represent 1st, 2nd, and 3rd positions. Map references **II E-16, II D-15, II D-13**

14th VERMONT INFANTRY
1st Corps, 3rd Div., 3rd Brig.

Raised: Counties of Addison, Rutland and Bennington.

Organized: 9 month regiment formed in Brattleboro, Vt. M.I. Oct. 21, 1862.

Commander: Col. William Thomas Nichols. b. Clarendon, Vt., Mar. 24, 1829. State Attorney for Rutland, Vt., 1859-1860. Pvt., 1st Vt. Infantry. M.O. Aug.

1861 to return to Rutland to represent the city in the state legislature. Col., 14th Vt., Oct. 21, 1862. M.O. July 30, 1863. Moved to Maywood, Ill., where he died April 10, 1882.

Number: On June 27, the regiment had 722 men.   **Loss:** 19 k., 67 wd., 21 m.

Monument: Hancock Ave. Map reference **II E-16**
   "July 2 & 3, 1863."

15th VERMONT INFANTRY
1st Corps, 3rd Div., 3rd Brig.

Raised: Counties of Caledonia, Orleans, Orange and Windsor.

Organized: 9 month regiment formed in Brattleboro, Vt. M.I. Oct. 22, 1862.

Commander: Col. Redfield Proctor. b. Proctorsville, Vt., June 1, 1831. Grad. Dartmouth College, N.Y., 1851. Grad. Albany Law School, N.Y., 1859. Lawyer in Boston. Returned to Vermont to enlist in 3rd Vt. as Lt. Maj., 5th Vt. Inf. Col., 15th Vt., Sept. 26, 1862. M.O. Aug. 5, 1863. Elected Governor of Vt., 1878; U.S. Senate, 1892; and appt. Secretary of War, 1889. d. Mar. 4, 1908.

Number: 637   **Loss:** None

Monument: None
   Arrived on the Battlefield p.m. on July 1. Took position on Cemetery Hill. Noon, July 2, sent to the rear to guard ammunition train.

16th VERMONT INFANTRY
1st Corps, 3rd Div., 3rd Brig.

Raised: Counties of Windsor and Windham.

Organized: 9 month regiment formed in Brattleboro, Vt. M.I. Oct. 23, 1862.

Commander: Col. Wheelock Graves Veazey. b. Brentwood, N.H., Dec. 5, 1835. Grad. Dartmouth College, N.H., 1859. Grad. Albany Law School, N.Y., 1860. Began practice in Springfield, Vt. Pvt., 3rd Vt. Infantry. Rose in rank to Lt. Col. before joining 16th Vt. as Col. to date Sept. 27, 1862. M.O. Aug. 10, 1863. Awarded Medal of Honor for gallantry at Gettysburg. Served in the state legislature. Appointed to State Supreme Court, 1879. d. Washington, D.C., Mar. 22, 1898. Buried Arlington National Cemetery.

Number: 715   **Loss:** 16 k., 102 wd., 1 m.

Monument: Hancock Ave. Map reference **II E-16**
   "Participated near this point in action of July 2nd. Picketed this line that night-held same as skirmishers until attacked by Pickett's Division, July 3rd. Rallied here and assaulted his flank to the right 400 yards-then changing front charged left flank of Wilcox's and Perry's brigades. At this point captured many hundred prisoners and two stands of colors."
   "The point to which the above inscription refers is south 58 degrees, west 1000 feet from this monument and near the northerly end of the Codori thicket. "

170

1st VERMONT CAVALRY
Cavalry Corps, 3rd Div., 1st Brig.

Raised: Counties of Franklin, Chittenden, Orleans, Lamoille, Washington, Addison, Rutland, Windsor, Bennington and Windham.

Organized: Camp Ethan Allen, Burlington, Vt. M.I. Nov. 19, 1861.

Commander: Lt. Col. Addison Webster Preston. b. Burke, Vt., Dec. 8, 1830. Attnd. Brown University, R.I. Farmer in Danville, Vt. Capt., Co. D, 1st Vt. Cavalry, Oct. 25, 1861. Wd. Sept. 20, 1862. Killed at Haw's Shop, Va., June 3, 1864.

Number: 687   **Loss:** 13 k., 25 wd., 27m.

Monuments: Slyder Field. Map reference **V I-7**

"In the Gettysburg Campaign, this regiment fought Stuart's cavalry at Hanover, Penn., June 30, and at Hunterstown, July 2; and on this field, July 3, led by Gen. Elon. J. Farnsworth, who fell near this spot, charged through the 1st Texas Infantry and to the line of Law's Brigade, receiving the fire of five Confederate Regiments and two batteries, and losing 67 men."

Confederate Ave—Figure of Maj. William Wells (1837-1892). Wells was awarded Medal of Honor for actions at Gettysburg. Map reference **V K-5**

7th WEST VIRGINIA INFANTRY
2nd Corps, 3rd Div., 1st Brig.

Raised: Counties of Hardy (note-Grant County formed from Hardy in 1866), Monongalia, Tyler, Marshall and Preston. Also Monroe County, Ohio and Greene County, Pennsylvania.

Organized: Camp Carlisle, Wheeling, W.V. M.I. July-Dec., 1861.

Commander: Lt. Col. Jonathan Hopkins Lockwood. b. Belmont County, Oh., 1808. Before the war lived in Moundsville, W.V. Member of the firm of Lockwood, Burley and Company conducting a milling, mining and merchandising business. Maj., 7th W.V., Dec. 20, 1861. Wd. Feb. 6, 1864 and May 12, 1864. M.O. Dec. 6, 1864. d. Moundsville, Mar. 28, 1892.

Number: 319   **Loss:** 5 k., 41 wd., 1 m.

Monument: East Cemetery Hill. Map reference **II K-3**

"At dusk July 2nd Carroll's Brigade was ordered by General Hancock to this point. 'On arriving there we found the Battery about to be taken charge of by the enemy who were in large force. Whereupon we immediately charged on the enemy and succeeded in completely routing their entire force and driving them beyond our lines.'"

Markers: Wainwright Ave., Cemetery Hill and near Cyclorama Center. Map references **II L-3, II K-3, II H-9**

1st WEST VIRGINIA ARTILLERY, BATTERY C
Artillery Reserve, 3rd Volunteer Brigade

Other Names: "Pierpont Battery"

Raised: Washington County, Ohio. The battery was formed and accepted by the governor of West Virginia.

Organized: Camp Carlisle, Wheeling, W.V. M.I. Mar. 30, 1862. Originally men made up Co. B, 18th Oh. Infantry (3 mos.). The battery was formed and offered to the Governor of West Virginia, who accepted it.

Commander: Capt. Wallace Hill. b. Marietta, Oh., Jan. 22, 1839. Farmer in Marietta, Oh. 2 Lt., Co. B, 18th Oh. Inf. (3 mos.). Lt., Battery C, Sept. 1, 1861. M.O. June 28, 1865. d. Sidney, Ohio, April 12, 1895.

Number: 4 10-pdr. Parrotts. 124 men.   **Loss:** 2 k., 2 wd.

Monument: National Cemetery. Map reference **II I-6**

Excelsior Brigade (70, 71, 72, 73, 74 New York Infantry), Sickles Avenue.

1st WEST VIRGINIA CAVALRY
10 Companies
Cavalry Corps, 3rd Div., 1st Brig.

Raised: Counties of Wirt, Wood, Ohio, Mason and Wayne in West Virginia. Penn. counties of Washington and Fayette. Also Ohio counties of Belmont and Jefferson.

Organized: Camp Carlisle, Wheeling, W.V., Clarksburg, W.V. and Morgantown, W.V. M.I. July-Oct., 1861. At Gettysburg companies A & I were not with regiment.

Commander: Col. Nathaniel Pendleton Richmond. b. Indianapolis, Ind., July 26, 1833. Attnd. Brown University, R.I. Lawyer and farmer in Kokomo, Indiana. 2 Lt., 13th Indiana Infantry. Lt. Col., 1st West Virginia, Sept. 7, 1861. M.O. Nov. 11, 1863. Member of Indiana state legislature and mayor of Kokomo. d. Malvern, Arkansas, June 28, 1919.

Number: 436   **Loss:** 4 k., 4 wd., 4 m.

Monument: Taneytown Road. Map reference II J-16

3rd WEST VIRGINIA CAVALRY
2 Companies, AC
Cavalry Corps, 1st Div., 2nd Brig.

Raised: No one predominant location.

Organized: A-Wheeling, W.V. M.I. Dec. 23, 1861. C-Brandonville, W.V. M.I. Oct. 1, 1861. Companies and squadrons not combined into a regiment until 1864.

Commander: Capt. Seymour Beach Conger. b. Plymouth, Ohio, Sept. 25, 1825. Farmer near Lexington, Oh. Recruited Company A. Capt., Co. A, 3rd W.V. Cav., Nov. 22, 1862. Killed as Maj., Aug. 7, 1864, near Moorefield, W.V. Buried Arlington National Cemetery.

Number: 59   **Loss:** 4 m.

Monument: Buford Ave. Map reference I C-5

2nd WISCONSIN INFANTRY
1st Corps, 1st Div., 1st Brig.

Raised: Counties of LaCrosse, Grant, Rock, Winnebago, Racine, Columbia and Iowa.

Organized: Camp Randall, Madison, Wisc. M.I. June 11, 1861.

Commanders: Col. Lucius Fairchild. b. Franklin Mills (modern Kent), Ohio, Dec. 27, 1831. Traveled to California gold fields in 1849 and 1855. Militia service. Lawyer in Madison, Wisc. Pvt., 1st Wisc. Infantry. Lt. Col., 2nd Wisc., June 20, 1861. Also held commission in Regular service. Wd. July 1, 1863. Left arm amputated. Resigned volunteer service Aug. 20, 1863. Resigned Regular service Oct. 19, 1863. Appt. Brig. Gen. of Volunteers, Oct. 19, 1863. Resigned Nov. 2, 1863 to accept position as Secretary of State of Wisconsin. Elected Governor of Wisconsin, 1866-72. d. Madison, Wisc., May 23, 1896. When Fairchild wounded, Maj. John Mansfield commanded regiment. b. New York State, 1822 or 1823. Lawyer in Portage, Wisc. Capt., Co. G, 2nd Wisc., June 11, 1861. Wd. July 1, 1863 and May 5, 1864. M.O;. Lt. Col., Aug. 24, 1864. Service in Veteran Reserve Corps. M.O. Nov. 20, 1866. Served as Lt. Governor of the California. d. Los Angeles, California, May 6, 1896. When Mansfield wd., Capt. George Henry Otis took command. b. Essex County, N.Y., Oct. 10, 1838. Printer in Mineral Point, Wisc. Pvt., Co. I, 2nd Wisc., April 26, 1861. M.O. June 24, 1864. Brief service in 8th U.S. Infantry. d. Larimer County Hospital, Fort Collins, Colorado, Jan. 28, 1931.

Number: 302   **Loss:** 26 k., 155 wd., 52 m.

Monument: Meredith Ave. Map reference I H-2

"July 1st 1863."

Markers: Slocum Ave. Map reference III C-3

3rd WISCONSIN INFANTRY
12th Corps, 1st Div., 3rd Brig.

Raised: Counties of Dodge, Lafayette, Winnebago and Grant.

Organized: "Armory Block" in Fond du Lac, Wisc. M.I. June 28, 1861.

Commander: Col. William Hawley. b. Lockport, N.Y., Aug. 19, 1824. Mexican War service. Furniture maker and dealer in Madison, Wisc. Capt., Co. K, 3rd Wisc., April 24, 1861. Wd. Aug. 9, 1862, May 1, 1863 and May 25, 1864 M.O. July 18, 1865. d. Buffalo, N.Y. in Regular service, Jan. 15, 1873.

Number: 260   **Loss:** 2 k., 8 wd.

Monument: Colgrove Ave. Map reference III H-11

"This regment went into position on this part of the line on the evening of the 1st. It moved to the left to reinforce the 3rd Corps. Returned to this position the same night and remained until the morning of July 5."

5th WISCONSIN INFANTRY
6th Corps, 1st Div., 3rd Brig.

Raised: Counties of Milwaukee, Manitowoc, Green Lake, Richland and Dunn.
Organized: Camp Randall, Madison, Wisc. M.I. July 13, 1861.
Commander: Col. Thomas Scott Allen. b. Allegany County, N.Y., July 26, 1825. Attnd. Oberlin College, Oh. Member Wisconsin state legislature. Member of militia. Before the war clerk in land office in Madison, Wisconsin. Capt., 2nd Wisc. Inf. Wd. Sept. 17, 1862. Col., 5th Wisc., Jan. 26, 1863. Wd. Nov. 7, 1863. M.O. Aug. 2, 1864. Wisconsin Secretary of State. d. Oshkosh, Wisconsin, Dec. 12, 1905.
Number: 491 **Loss:** None
Monument: Howe Ave. Map reference **V L-18**
"This regiment moved from centre to this point early July 3rd to resist threatened attack on this flank. Moved hastily back in the afternoon to assist in repelling attack on the centre and later took position on the crest of Big Round Top."

6th WISCONSIN INFANTRY
1st Corps, 1st Div., 1st Brig.

Raised: Counties of Sauk, Pierce, Crawford, Milwaukee, Rock and Vernon.
Organized: Camp Randall, Madison, Wisc. M.I. July 16, 1861.
Commander: Lt. Col. Rufus R. Dawes. b. Malta, Ohio, July 4, 1838. Great grandson of William Dawes, Jr. who was with Paul Revere on his famous ride. Father of Vice President Charles G. Dawes. Grad Marietta College, Oh., 1860. In Juneau County Wisconsin on business when the war began. Capt., Co. K, 6th Wisc., July 16, 1861. M.O. Col., Aug. 10, 1864. Elected to U.S. Congress. d. Marietta, Oh., Aug. 1, 1899.
Number: 340 **Loss:** 30 k., 116 wd., 22 m.
Monument: Reynolds Ave. Map reference **I G-5**
"On July 2 & 3, this Regt. lay on Culp's Hill. On the evening of the 2nd it moved to support of Greene's Brigade and assisted to repulse Johnson's Division."
"In the charge made on this R.R. Cut the 2nd Miss. Regt., officers, men and battle flag, surrendered to the 6th Wisc."
Marker: Slocum Ave. Map reference **III D-2**

7th WISCONSIN INFANTRY
1st Corps, 1st Div., 1st Brig.

Raised: Counties of Columbia, Grant, Dane, Waushara and Marquette.
Organized: Camp Randall, Madison, Wisc. M.I. Sept. 13, 1861.
Commanders: Col. William Wallace Robinson. b. Fairhaven, Vt., Dec. 14, 1819. Mexican War service. Farmer in Sparta, Wisconsin. Lt. Col., 7th Wisc., Aug. 18, 1861. Wd. Aug. 28, 1862. M.O. July 9, 1864. d. Seattle, Washington, April

27, 1903. When Robinson elevated to brigade command, Maj. Mark Finnicum led the regiment. b. Ohio, Dec. 25, 1823. Merchant in Fenimore, Wisc. Capt., Co. H, 7th Wisc., July 20, 1861. Wd. April 29, 1863. M.O. Lt. Col., Dec. 17, 1864. d. Pulaski County, Kentucky, Nov. 4, 1912.

Number: 343 **Loss:** 21 k., 105 wd., 52 m.

Monument: Meredith Ave. Map reference **I H-2**

"This monument marks one of the advanced positions of the Regt. in battle July 1st 1863. Position of Regt. July 2 & 3, indicated by stone marker on Culp's Hill."

Marker: Slocum Ave. Map reference **III C-3**

26th WISCONSIN INFANTRY
11th Corps, 3rd Div., 2nd Brig.

Raised: Counties of Milwaukee, Fond du Lac, Manitowoc and Washington.

Organized: Camp Sigel, Milwaukee, Wisconsin. M.I. Sept. 17, 1862.

Commanders: Lt. Col. Hans Boebel. b. Bavaria, Mar. 10, 1829. Took part in revolutions in Germany, eventually coming to the U.S. in 1852. Printer in Milwaukee. 2 Lt., 5th Wisc. Inf. 2 Lt., Co. H, 26th Wisc., Aug. 14, 1862. Wd. July 1, 1863. M.O. May 28, 1864. d. Milwaukee, Aug. 24, 1902. When Boebel wounded, Capt. John William Fuchs took command. b. Zulpich, Prussia, Sept. 21, 1828. Veterinary surgeon in Milwaukee. Corpl., 5th Wisc. Inf. Lt., Co. C, 26th Wisc., Sept. 17, 1862. M.O. June 7, 1865. d. Milwaukee, Jan. 30, 1870.

Number: 516 **Loss:** 26 k., 129 wd., 62 m.

Monument: Howard Ave. Map reference **I D-14**

"July 1, 1863. On Cemetery Hill July 2 and 3."

4th New York Battery, Sickles Avenue, Devil's Den.

INDEX

Baxter, Brig. Gen. Henry, 2nd Brig., 2nd Div., 1st Corps.

Beardsley, Capt. John D., 10th Me. Inf.

Beardsley, Maj. William E., 6th N.Y. Cav.

Beaumont, Maj. Myron H., 1st N.J. Cav.

Beaver County, Pennsylvania
Penn. Inf., 9th Penn. Res., 10th Penn. Res., 140th.
Penn Cav., 17th.

Belknap County, New Hampshire
N.H. Inf., 12th.

Belmont County, Ohio
Oh. Inf., 25th, 61st.
West Virginia Cav., 1st.

Bennington County, Vermont
Vt. Inf., 2nd, 4th, 14th.
Vt. Cav., 1st.

Bentley, Lt. Col. Richard C., 63rd N.Y. Inf.

Berdan, Col. Hiram, 1st U.S.S.S.

Berks, County, Pennsyivania
Penn. Inf., 46th, 88th, 93rd, 151st.
Penn. Cav., 1st, 6th.
U.S. Regular Art., 5th (K).

Berkshire County, Massachusetts
Mass. Inf., 10th, 37th.

Beveridge, Maj. John L., 8th Ill. Cav.

Biddle, Maj. Alexander, 121st Penn. Inf.

Biddle, Col. Chapman, 121st Penn.

Biddle, Col. George H., 95th N.Y. Inf.

Bidwell, Col. Daniel D., 49th N.Y. Inf.

Bigelow, Capt. John, 9th Mass. Art.

Bingham, Col. Daniel G., 64th N.Y. Inf.

Birney, Maj. Gen. David B., 1st Div., 3rd Corps.

Blair County, Pennsylvania
Penn. Inf., 62nd, 110th.
U.S. Regular Art., 5th (K).

Bodine, Maj. Robert L., 26th Penn. Inf.

Boebel, Lt. Col. Hans, 26th Wisc. Inf.

Bootes, Capt. Levi C., 6th U.S. Inf.

Both, Lt. Ernst, 54th N.Y. Inf.

Bowen, Capt. Edward R., 114th Penn. Inf.

Boynton, Maj. Joseph J., 13th Vt. Inf.

Bradford County, Pennsyivania
Penn. Inf., 5th Penn. Res., 6th Penn. Res., 12th Penn. Res., 57th, 106th, 141st.
Penn. Cav., 17th.

Bradley, Maj. Leman W., 64th N.Y. Inf.

Brady, Maj. Allen G., 17th Conn. Inf.

Branch County, Michigan
Mich. Cav., 5th.

Breck, Lt. George, 1st N.Y. Art., Battery L.

Brewster, Col. William R., 2nd Brig., 2nd Div., 3rd Corps.

Brinton, Lt. Col. William P., 18th Penn. Cav.

Bristol County, Massachusetts
Mass. Inf., 7th, 18th, 22nd, 33rd.
Mass. Art., 5th.

Broady, Lt. Col. Knut O., 61st N.Y. Inf.

Brooke, Col. John R., 4th Brig., 1st Div., 2nd Corps.

Broome County, New York
N.Y. Inf., 137th.

Brown (Barnes), Col. Henry W., 3rd N.J. Inf.

Brown, Col. Hiram L., 145th Penn. Inf.

Brown, Col. Philip P., Jr., 157th N.Y. Inf.

Brown, Lt. Thomas F., 1st R.I. Art., Batt. B.

Bucklyn, Lt. John K., 1st R.I. Art., Batt. E.

Buford, Brig. Gen. John, 1st Div., Cavalry Corps.

Bull, Lt. Col. James M.,126th N.Y. Inf.

Burbank, Col. Sidney, 2nd Brig., 2nd Div., 5th Corps.

Burke, Capt. Denis F., 88th N.Y. Inf.

Burling, Col. George C., 3rd Brig., 2nd Div., 3rd Corps.

Burlington County, New Jersey
N.J. Inf., 3rd, 4th, 5th, 6th, 12th.
N.J. Cav., 1st.

Burnham, Col. Hiram, 6th Me. Inf.

Burns, Maj. Michael W., 73rd N.Y. Inf.

Butler County, Pennsylvania
Penn. Inf., 11th Penn. Res.

Butler, Lt. John H., 2nd U.S. Art., Batt. G.

Byrnes, Col. Richard, 28th Mass. Inf.

Cain, Lt. Col. John H., 155th Penn. Inf.

Caldwell, Brig. Gen. John C., 1st Div., 2nd Corps.

Caledonia County, Vermont
Vt. Inf., 3rd, 4th, 15th.

Calef, Lt. John H., 2nd U.S. Art., Battery A.

Cambria County, Pennsylvania
Penn. Inf., 11th Penn. Res., 115th.
Penn. Cav., 18th.

Camden County, New Jersey
N.J. Inf., 1st, 3rd, 4th, 6th, 12th.

Cameron County, Pennsylvania
  Penn. Inf., 13th Penn. Res.
Candy, Col. Charles, 1st Brig., 2nd Div.,
  12th Corps.
Cantador, Lt. Col. Lorenz, 27th Penn. Inf.
Carbon County, Pennsylvania
  Penn Inf., 11th, 28th, 13th Penn. Res.,
  81st.
Carman, Col. Ezra A., 13th N.J. Inf.
Caroline County, Maryland
  Md. Inf., 1st Md. East. Shore.
Carpenter, Lt. Col. Leonard W., 4th Oh. Inf.
Carr, Brig Gen. Joseph B., 1st Brig., 2nd
  Div., 3rd Corps.
Carroll County, New Hampshire
  N.H. Inf., 5th, 12th.
Carroll, Lt. Col. Edward, 95th Penn. Inf.
Carroll, Col. Samuel S., 1st Brig., 3rd Div.,
  2nd Corps.
Cass County, Indiana
  Ind. Inf., 20th.
  U.S. Regular Inf., 12th.
Cattaraugus County, New York
  N.Y. Inf., 64th, 71st, 154th.
  N.Y. Cav., 9th.
Cayuga County, New York
  N.Y. Inf., 111th.
  N.Y. Art., 1st.
Cavada, Lt. Col. Frederick F., 114th Penn.
  Inf.
Cecil County, Maryland Delaware Inf.,
  2nd.
Centre County, Pennsylvania
  Penn. Inf., 5th Penn. Res., 49th, 56th,
  148th.
  Penn. Cav., 1st, 2nd.
Chamberlain, Col. Joshua L., 20th Me. Inf.
Champaign County, Ohio
  Oh. Inf., 66th.
Chapman, Lt. Col. Alford B., 57th N.Y. Inf.
Chapman, Col. George H., 3rd Ind. Cav.
Chautauqua County, New York
  N.Y. Inf., 49th, 72nd, 154th.
  N.Y. Cav., 9th.
Chemung County, New York
  N.Y. Inf., 86th, 107th.
  N.Y. Cav., 10th.
  U.S. Regular Inf., 14th.
Chenango County, New York

N.Y. Cav., 8th, 10th.
Cheshire County, New Hampshire
  N.H. Inf., 2nd.
Chester County, Pennsylvania
  Penn. Inf., 1st Penn. Res., 13th Penn.
  Res., 49th.
Chittenden County, Vermont
  Vt. Inf., 2nd, 5th, 6th, 13th.
  Vt. Cav., 1st.
Christman, Capt. Charles H., 2nd Del. Inf.
Clarion County, Pennsylvania
  Penn. Inf., 10th Penn. Res., 62nd, 63rd,
  148th, 155th.
Clark, Capt. Adoniram J., 1st N.J. Art.,
  Battery B.
Clearfield County, Pennsylvania
  Penn. Inf., 5th Penn. Res., 13th Penn.
  Res., 149th.
Clinton County, Pennsylvania
  Penn. Inf., 11th.
  Penn. Cav., 1st.
Clinton, Capt. William, 10th U.S. Inf.
Coates, Capt. Henry C., 1st Minn. Inf.
Cobham, Col. George A., Jr., 2nd Brig., 2nd
  Div., 12th Corps.
Coburn, Maj. James H., 27th Conn. Inf.
Colgrove, Col. Silas, 27th Ind. Inf.
Collier, Col. Frederick H., 139th Penn. Inf.
Columbia County, New York
  N.Y. Cav., 6th.
Columbia County, Pennsylvania
  Penn. Inf., 6th Penn. Res.
Columbia County, Wisconsin
  Wisc. Inf., 2nd, 7th.
Colvill, Col. William, Jr., 1st Minn. Inf.
Conger, Capt. Seymour B., 3rd W.V. Cav.
Conner, Lt. Col. Freeman, 44th N.Y. Inf.
Connor, Lt. Col. Selden, 7th Me. Inf.
Cook County, Illinois
  Ill. Inf., 82nd.
  Ill. Cav., 8th, 12th.
  N.Y. Art., 1st (B), 1st (G).
  U.S. Regular Art., 4th (G).
Cook, Maj. John E., 76th N.Y. Inf.
Coons, Col. John, 14th, Ind. Inf.
Cooper, Maj. Frederick, 7th N.J. Inf.
Cooper, Capt. James H., 1st Penn. Art.,
  Battery B.
Coos County, New Hampshire

179

N.H. Inf., 5th.
Me. Art., 5th.
Cortland County, New York
N.Y. Inf., 76th, 157th.
N.Y. Cav., 10th.
Coster, Col. Charles R., 1st Brig., 2nd Div., 11th Corps.
Coulter, Col. Richard, 11th Penn. Inf.
Cowan, Capt. Andrew, 1st N.Y. Art.
Craig, Col. Calvin A., 105th Penn. Inf.
Crandell, Lt. Col. Levin, 125th N.Y. Inf.
Crane, Col. Nirom M., 107th N.Y. Inf.
Crawford County, Ohio
Oh. Inf., 8th.
Crawford County, Pennsylvania
Penn. Inf., 9th Penn. Res., 10th Penn. Res., 57th, 83rd, 111th, 145th, 150th.
Penn. Cav., 2nd, 18th.
Crawford County, Wisconsin
Wisc. Inf., 6th.
Crawford, Brig. Gen. Samuel W., 3rd Div., 5th Corps.
Creighton, Col. William R., 7th Oh. Inf.
Cross, Col. Edward E., 1st Brig., 1st Div., 2nd Corps.
Cross, Col. Nelson, 67th N.Y. Inf.
Cumberland County, Maine
Me. Inf., 5th, 10th, 17th.
Me. Cav., 1st.
U.S. Regular Inf., 17th.
Cumberland County, New Jersey
N.J. Inf., 3rd, 12th.
Cumberland County, Pennsylvania
Penn. Inf., 11th, 1st Penn. Res., 107th.
Penn. Cav., 3rd, 17th.
Cummins, Lt. Col. Francis M., 124th N.Y. Inf.
Cummins, Col. Robert P., 142nd Penn. Inf.
Cunningham, Lt. Col. Henry W., 19th Me. Inf.
Curry, Lt. Col. William L., 106th Penn. Inf.
Curtis, Lt. Col. Greely S., 1st Mass. Cav.
Curtiss, Maj. Sylvanus W., 7th Mich. Inf.
Cushing, Lt. Alonzo H., 4th U.S. Art., Battery A.
Custer, Brig. Gen. George A., 2nd Brig., 3rd Div. Cavalry Corps.
Cutler, Brig. Gen. Lysander, 2nd Brig., 1st Div., 1st Corps.

Cuyahoga County, Ohio
Oh. Inf., 7th, 8th, 107th.
Oh. Art., 1st (K).

Dakota, County, Minnesota
Minn. Inf., 1st.
Dana, Col. Edmund L., 143rd Penn. Inf.
Dane County, Wisconsin
Wisc. Inf., 7th.
Daniels, Capt. Jabez J., 9th Mich. Art.
Danks, Maj. John A., 63rd Penn. Inf.
Dare, Lt. Col. George, 5th Penn. Res. Inf.
Darrow, Capt. John, 82nd, N.Y. Inf.
Dauphin County, Pennsylvania
Penn. Inf., 6th Penn. Res., 12th Penn. Res., 46th, 107th, 147th.
Penn. Cav., 18th.
Daviess County, Indiana
Ind. Inf., 27th.
Davis, Capt. Milton S., 68th Penn. Inf.
Davis, Capt. William, 69th Penn. Inf.
Dawes, Lt. Col. Rufus R., 6th Wisc. Inf.
Day, Col. Hannibal, 1st Brig., 2nd Div., 5th Corp.
Dearborn County, Indiana
Ind. Inf., 7th.
Ind. Cav., 3rd.
Decatur County, Indiana
Ind. Inf., 7th.
Deems, Lt. Col. James M., 1st Md. Cav.
Defiance County, Ohio
Oh. Inf., 107th.
DeKalb County, Illinois
Ill. Cav., 8th.
Delaware County, Indiana
Ind. Inf., 19th.
Delaware County, New York
N.Y. Inf., 72nd.
Delaware County, Ohio
Oh. Inf., 4th, 66th.
Delaware County, Pennsylvania
Penn. Inf., 1st Penn. Res., 119th.
Dent, Lt. John T., 1st Del. Inf.
Des Moines County, Iowa
U.S. Regular Inf., 11th.
De Trobriand, Col. P. Regis. 3rd Brig., 1st Div., 3rd Corps.
Devereux, Col. Arthur F., 19th Mass. Inf.
Devin, Col. Thomas C., 2nd Brig., 1st Div.,

Cavalry Corps.

Dilger, Capt. Hubert, 1st Oh. Art., Battery I.

Dobke, Lt. Col. Adolphus, 45th N.Y. Inf.

Dodge County, Wisconsin
Wisc. Inf., 3rd.

Donovan, Capt. Matthew, 16th Mass. Inf.

Dorchester County, Maryland
Md. Inf., 1st Md. Eastern Shore.

Doster, Lt. Col. William E., 4th Penn. Cav.

Doubleday, Maj. Gen. Abner, 3rd Div., 1st Corps.

Dow, Lt. Edwin B., 6th Me. Art.

Du Bois County, Indiana
Ind. Inf., 27th.

Dubuque County, Iowa
U.S. Regular Inf., 12th.

Dunn County, Wisconsin
Wisc. Inf., 5th.

Dunn, Capt. Thomas S., 12th U.S. Inf.

Dunne, Maj. John P., 115th Penn. Inf.

DuPage County, Illinois
Ill. Cav., 8th.

Dutchess County, New York
N.Y. Inf., 57th, 150th.

Duvall, Capt. Robert E., Purnell Legion Md. Cav.

Dwight, Col. Walton, 149th, Penn. Inf.

Eakin, Lt. Chandler P., 1st U.S. Art., Battery H.

Eaton County, Michigan
Mich. Cav., 7th.

Edgell, Capt. Frederick M., 1st N.H. Art.

Edwards, Capt. Albert M., 24th Mich. Inf.

Edwards, Col. Clark S., 5th Me. Inf.

Edwards, Col. Oliver, 37th Mass. Inf.

Egan, Col. Thomas W., 40th N.Y. Inf.

Elder, Lt. Samuel S., 4th U.S. Art., Batt. E.

Elk County, Pennsylvania
Penn. Inf., 13th Penn. Res.

Elkhart County, Indiana
Ind. Inf., 19th.

Ellis, Col. Augustus Van Horne, 124th N.Y. Inf.

Ellis, Maj. Theodore G., 14th Conn. Inf.

Ellmaker, Col. Peter C., 119th Penn. Inf.

Ent, Lt. Col. Wellington H., 6th Penn. Res.

Erie County, New York

N.Y. Inf., 44th, 49th, 78th. N.Y. Art., 1st N.Y. (I).

N.Y. Cav., 10th.

U.S. Regular Inf., 17th.

U.S. Regular Art., 5th (F).

Erie County, Ohio
Oh. Inf., 8th, 55th, 107th.

Erie County, Pennsylvania
Penn. Inf., 83rd, 111th, 145th.
Penn. Cav., 16th.

Ernst, Lt. Col. Louis, 140th N.Y. Inf.

Essex County, Massachusetts
Mass. Inf., 2nd, 9th, 11th, 12th, 19th, 22nd, 32nd, Mass. Sharpshooters; 1st Co.
Mass. Cav., 1st.

Essex County, New Jersey
N.J. Inf., 2nd, 5th, 7th, 8th, 11th, 13th.
N.J. Art., 1st (B).
N.J. Cav.,1st. N.Y. Inf., 41st, 70th, 71st, 72nd.
U.S. Regular Inf., 12th.

Essex County, New York
N.Y. Inf., 77th.
N.Y. Cav., 5th.

Essex County, Vermont Vt. Inf., 3rd.

Eustis, Col. Henry L., 2nd Brig., 3rd Div., 6th Corps.

Ewing, Maj. Charles, 4th N.J. Inf.

Fairchild, Col. Lucius, 2nd Wisc. Inf.

Fairfield County, Connecticut
Conn. Inf., 5th, 14th, 17th.
Conn. Art., 2nd.

Farnsworth, Brig. Gen. Elon J., 1st Brig., 3rd Div. Cavalry Corps.

Farnum, Col. John E., 70th N.Y.

Fayette County, Indiana Ind. Cav., 3rd.

Fayette County, Ohio
Oh. Cav., 1st.

Fayette County, Pennsylvania
Penn. Inf., 11th Penn. Res., 142nd.
Penn. Cav., 1st, 16th.
W.V. Cav., 1st.

Fesler, Lt. Col. John R., 27th Ind. Inf.

Finnicum Maj. Mark, 7th Wisc. Inf.

Fisher, Col. Joseph W., 3rd Brig., 3rd Div., 5th Corps.

Fitzhugh, Capt. Robert H., 1st N.Y. Art.,

Battery K.

Floyd Jones, Maj. Delancey, 11th U.S. Inf.

Flynn, Capt. John H., 28th, Penn. Inf.

Foerster, Lt. Hermann, 8th N.Y. Inf.

Fond du Lac County, Wisconsin
Wisc. Inf., 26th.

Forest County, Pennsylvania
Penn. Inf., 83rd.

Fountain County, Indiana
Ind. Inf., 20th.

Foust, Maj. Benezet F., 88th Penn. Inf.

Fowler, Lt. Col. Douglas, 17th Conn. Inf.

Fowler, Col. Edward B., 84th N.Y. Inf.

Fox, Capt. George B., 75th Oh. Inf.

Francine, Col. Louis R., 7th N.J. Inf.

Franklin County, Maine
Me. Inf., 16th, 17th.
Me. Cav., 1st.

Franklin County, Massachusetts
Mass. Inf., 10th.

Franklin County, New York
N.Y. Inf., 60th.

Franklin County, Ohio
U.S. Regular Cav., 1st.

Franklin County, Pennsylvania
Penn. Inf., 6th Penn. Res., 12th Penn.
Res., 107th.
Penn. Cav., 16th, 17th.

Franklin County, Vermont
Vt. Inf., 2nd, 5th, 6th, 13th.
Vt. Cav., 1st.

Fraser, Lt. Col. John, 140th Penn. Inf.

Frederick County, Maryland
Md. Inf., 1st Pot. Home Brigade.

Freeborn, 2 Lt. Benjamin; 1st R.I. Art., Battery E.

Freedley, Capt. Henry W., 3rd U.S. Inf.

French, Lt. Col. Winsor B., 77th N.Y. Inf.

Freudenberg, Lt. Col. Charles G., 52nd N.Y. Inf.

Frueauff, Maj. John F., 153rd Penn. Inf.

Fuchs, Capt. John W., 26th Wisc. Inf.

Fuger, Sergt. William F., 4th U.S. Art., Battery A.

Fulton County, New York
N.Y. Inf., 77th.
N.Y. Cav., 10th.

Gambee, Col. Charles B., 55th Oh. Inf.

Gamble, Col. William, 1st Brig., 1st Div., Cavalry Corps.

Garrard, Col. Kenner, 146th N.Y. Inf.

Gates, Col. Theodore B., 80th N.Y. Inf.

Geary, Brig. Gen. John W., 2nd Div., 12th Corps.

Gibbon, Brig. Gen. John, 2nd Div., 2nd Corps.

Gibbs, Capt. Frank C., 1st Oh. Art., Battery L.

Giddings, Maj. Grotius R., 14th U.S. Inf.

Gifford, Capt. Henry J., 33rd N.Y. Inf.

Gilkyson, Lt. Col. Stephen R., 6th N.H. Inf.

Gimber, Capt. Frederick L., 109th Penn. Inf.

Glenn, Capt. James, 149th Penn. Inf.

Glenn, Lt. Col. John F., 23rd Penn. Inf.

Gloucester County, New Jersey
N.J. Inf., 3rd, 12th.

Godard, Col. Abel, 60th N.Y. Inf.

Godfrey, Capt. Thomas C., 5th N.J. Inf.

Goodhue County, Minnesota
Minn. Inf., 1st.

Grafton County, New Hampshire
N.H. Inf., 5th, 12th.

Graham, Brig. Gen. Charles K., 1st Brig., 1st Div., 3rd Corps.

Graham, Capt. William M., 1st U.S. Art., Battery K.

Grant County, West Virginia
see Hardy County.

Grant County, Wisconsin
Wisc. Inf., 2nd, 3rd, 7th.

Grant, Col. Lewis A., 2nd Brig., 2nd Div., 6th Corps.

Gray, Col. George, 6th Mich. Cav.

Green Lake County, Wisconsin
Wisc. Inf., 5th.

Greene County, Indiana
Ind. Inf., 14th.

Greene County, New York
N.Y. Inf., 120th.

Greene County, Pennsylvania
Penn. Inf., 140th.
Penn. Cav., 1st, 18th.
W.V. Inf., 7th.

Greene, Brig. Gen. George S., 3rd Brig., 2nd Div., 12th Corps.

Greene, Lt. Col. James D., 17th U.S. Inf.

Gregg, Brig. Gen., David McM., 2nd Div.,

Cavalry Corps.

Gregg, Col. J. Irvin 3rd Brig., 2nd Div., Cavalry Corps.

Grover, Maj. Andrew J., 76th N.Y.

Grover, Col. Ira G., 7th Ind. Inf.

Grumbach, Capt. Nicholas, Jr., 149th N.Y. Inf.

Guiney, Col. Patrick R., 9th Mass. Inf.

Gwyn, Lt. Col. James, 118th Penn. Inf.

Haines, Capt. Benjamin F., 11th Penn. Inf.

Hall, Capt. James A., 2nd Me. Art.

Hall, Col. Norman J., 3rd Brig., 2nd Div., 2nd Corps.

Hamblin, Col. Joseph E., 65th N.Y. Inf.

Hamilton County, Ohio
Oh. Inf., 5th, 61st, 75th.
Oh. Art., 1st (I).
U.S. Regular Art., 1st (H), 2nd (A).
2nd (B&L consolidated), 2nd (D), 2nd (M).
U.S. Regular Cav., 1st, 5th.

Hamilton, Lt. Col. Theodore B., 62nd N.Y. Inf.

Hammell, Lt. Col. John S., 66th N.Y. Inf.

Hammond, Maj. John, 5th N.Y. Cav.

Hampden County, Massachusetts
Mass. Inf., 10th, 37th.
Mass. Cav., 1st.

Hampshire County, Massachusetts
Mass. Inf., 10th, 37th.

Hancock County, Maine
Me. Inf., 6th.

Hancock, Capt. David P., 7th U.S. Inf.

Hancock, Maj. Gen. Winfield S., 2nd Corps.

Hapgood, Lt. Col. Charles E., 5th N.H. Inf.

Hardin County, Ohio
Oh. Inf., 4th, 82nd.

Hardin, Col. Martin D., 12th Penn. Res. Inf.

Hardy County, West Virginia
W.V. Inf., 7th.

Harford County, Maryland
Md. Cav., Purnell Legion.

Harlow, Lt. Col. Franklin P., 7th Mass. Inf.

Harn, Capt. William A., 3rd N.Y. Art.

Harney, Maj. George, 147th N.Y. Inf.

Harris, Col. Andrew L., 75th Oh. Inf.

Harris, Lt. Col. Edward P., 1st Del. Inf.

Harrison County, Indiana

Ind. Cav., 3rd.

Harrow, Brig. Gen. William, 1st Brig., 2nd Div., 2nd Corps.

Hart, Capt. Patrick, 15th N.Y. Art.

Hartford County, Connecticut
Conn. Inf., 5th, 14th, 20th.

Hartshorne, Maj. William R., 13th Penn. Res. Inf.

Haseltine, Maj. James H., 6th Penn. Cav.

Hawley, Col. William, 3rd Wisc. Inf.

Hayes, Capt. Edward, 29th Oh. Inf.

Hayes, Col. Joseph, 18th Mass. Inf.

Hays, Brig. Gen. Alexander, 3rd Div., 2nd Corps.

Hazard, Capt. John G., 2nd Corps Art.

Hazlett, Lt. Charles E., 5th U.S. Art., Battery D.

Heath, Col. Francis E., 19th Me. Inf.

Heaton, Lt. Edward, 2nd U.S. Art., Battery B + L.

Heckman, Capt. Lewis, 1st Oh. Art., Battery K.

Hendricks County, Indiana
Ind. Inf., 7th.

Hennepin County, Minnesota
Minn. Inf., 1st.

Henry, Lt. Col. William, Jr, 1st N.J. Inf.

Herkimer County, New York
N.Y. Inf., 97th, 121st.

Hesser, Lt. Col. Theodore, 72nd Penn. Inf.

Higgins, Lt. Col. Benjamin L., 86th N.Y. Inf.

Highland County, Ohio
Oh. Inf., 73rd.

Hill, Maj. John T., 12th N.J. Inf.

Hill, Capt. Wallace, 1st W.V. Art., Batt. C.

Hillebrandt, Maj. Hugo, 39th N.Y. Inf.

Hillsborough County, New Hampshire
N.H. Inf., 2nd.
N.H. Art., 1st.

Hillsdale County, Michigan
Mich. Inf., 4th.
Mich. Art., 9th.

Hizar, Capt. Thomas B., 1st Del. Inf.

Hofmann, Col. John W., 56th Penn. Inf.

Holt, Lt. Col. Thomas, 74th N.Y. Inf.

Hopper, Maj. George F., 10th N.Y. Inf.

Howard, Maj. Gen. Oliver O., 11th Corps.

Howe, Brig. Gen. Albion P., 2nd Div., 6th Corps.

Hudson County, New Jersey
N.J. Inf., 1st, 5th, 6th, 7th, 8th, 11th, 13th.
N.J. Art., 1st. 1 (A).
N.J. Cav., 1st.
Huidekoper, Lt. Col. Henry S., 150th Penn. Inf.
Hulings, Lt. Col. Thomas M., 49th Penn. Inf.
Hull, Lt. Col. James C., 62nd Penn. Inf.
Humphreys, Brig. Gen. Andrew A., 2nd Div., 3rd Corps.
Hunterdon County, New Jersey
N.J. Inf., 6th, 8th, 15th.
Huntingdon County, Pennsylvania
Penn. Inf., 5th Penn. Res., 12th Penn. Res., 49th, 110th, 147th, 149th.
Huntington, Capt. James F., 3rd Volunteer Brig., Artillery Reserve.
Huron County, Ohio
Oh. Inf., 7th, 8th, 55th.
Huston, Lt. Col. James, 82nd N.Y. Inf.

Indiana County, Pennsylvania
Penn. Inf., 11th Penn. Res., 12th Penn. Res., 56th, 61st, 105th.
Ionia County, Michigan
Mich. Inf., 3rd, 16th.
Mich. Cav., 6th.
Iowa County, Wisconsin
Wisc. Inf., 2nd.
Ireland, Col. David, 137th N.Y. Inf.

Jackson County, Michigan
Mich. Inf., 1st.
Jackson, Lt. Col. Allan H., 134th N.Y. Inf.
Jackson, Col. Samuel M., 11th Penn. Res.
James, 2 Lt. Robert, 4th U.S. Art, Battery K.
Jefferson County, Indiana
Ind. Cav., 3rd.
Jefferson County, New York
N.Y. Inf., 94th.
N.Y. Art., 1st (C), 1st (D).
Jefferson County, Ohio
W.V. Cav., 1st.
Jefferson County, Pennsylvania
Penn. Inf., 11th Penn. Res., 62nd, 105th, 148th.
Jeffords, Col. Harrison H., 4th Mich. Inf.

Jenkins, Lt. Col. David T., 146th N.Y. Inf.
Jennings County, Indiana
Ind. Inf., 27th.
Johnson County, Indiana
Ind. Inf., 7th, 19th, 27th.
Jones, Lt. Col. David M., 110th Penn. Inf.
Jones, Lt. Col. Edward S., 3rd Penn. Cav.
Jones, Capt. George W., 150th Penn. Inf.
Jones, Capt. Noah, 1st Oh. Cav.
Joslin, Lt. Col. George C., 15th Mass. Inf.
Juniata County, Pennsylvania
Penn. Inf., 49th, 53rd, 151st.
Penn Cav., 1st, 16th.

Kalamazoo County, Michigan
Mich. Cav., 5th, 7th.
Kane County, Illinois
Ill. Cav., 8th.
Kane, Brig. Gen. Thomas L., 2nd Brig., 2nd Div., 12th Corps.
Kankakee County, Illinois
Ill. Cav., 12th.
Kellogg, Col. Josiah H., 17th Penn. Cav.
Kelly, Capt. Daniel F., 73rd Penn. Inf.
Kelly, Col. Patrick, 2nd Brig., 1st Div., 2nd Corps.
Kennebec County, Maine
Me. Inf., 3rd, 7th, 16th, 19th.
Me. Cav., 1st.
Kent County, Delaware
Del. Inf., 1st.
Kent County, Michigan
Mich. Inf., 3rd.
Mich. Cav., 6th.
Kent County, Rhode Island
R.I. Inf., 2nd.
Ketcham, Col. John H., 150th N.Y. Inf.
Kilpatrick, Brig. Gen. Judson, 3rd Div., Cavalry Corps.
Kings County, New York
N.Y. Inf., 67th, 73rd, 84th.
N.Y. Art., 5th.
Kinzie, Lt. David H., 5th U.S. Art., Battery K.
Knox County, Indiana
Ind. Inf., 14th.
Knox County, Maine
Me. Inf., 4th, 19th, 20th.
Me. Art., 2nd.

Inf.

Macomb County, Michigan
Mich. Inf., 5th.

Macy, Lt. Col. George N., 20th Mass. Inf.

Madill, Col. Henry J., 141st Penn. Inf.

Madison County, Indiana
Ind. Inf., 19th.

Madison County, New York
N.Y. Inf., 61st, 157th.
N.Y. Cav., Onedia Indpt. Co.

Mahier, Col. Francis, 75th Penn. Inf.

Mahoning County, Ohio
Oh. Inf., 7th.

Mallon, Col. James E., 42nd N.Y. Inf.

Maloney, 2 Lt. William, 25th Oh. Inf.

Manitowoc County, Wisconsin
Wisc. Inf., 5th, 26th.

Mann, Capt. Daniel P., Oneida N.Y. Independent Company of Cavalry.

Mann, Col. WilLiam D., 7th Mich. Cav.

Manning, Capt. Nathaniel J., 25th Oh. Inf.

Mansfield, Maj. John, 2nd Wisc. Inf.

Marion County, Indiana
Ind. Inf., 7th, 19th, 20th.
Ind. Cav., 1st.
U.S. Regular Inf., 11th.

Marion County, Ohio
Oh. Inf., 4th, 82nd.

Markell, Lt. Col. William L., 8th N.Y. Cav.

Marquette County, Wisconsin
Wisc. Inf., 7th.

Marshall County, Indiana
Ind. Inf., 20th.

Marshall County, West Virginia
W.V. Inf., 7th.

Marston, Capt. Daniel, 16th Me. Inf.

Martin County, Indiana
Ind. Inf., 14th.

Martin, Capt. Augustus P., 5th Corps Art.

Martin, Capt. Joseph W., 6th N.Y. Art.

Martin, Lt. Leonard, 5th U.S. Art., Battery F.

Martin, Capt. Luther, 11th N.J. Inf.

Mason County, West Virginia
W.V. Cav., 1st.

Mason, Capt. Julius W., 5th U.S. Cav.

Mason, 2 Lt. Philip D., 1st U.S. Art., Battery H.

Mass, Capt. Edmund A., 88th Penn. Inf.

Maulsby, Col. William P., Sr., 1st Md. Potomac Home Brig.

McAllister, Col. Robert, 11th N.J. Inf.

McCalmont, Lt. Col. Alfred B., 142nd Penn. Inf.

McCandless, Col. William, 1st Brig., 3rd Div., 5th Corps.

McCartney, Capt. William H., 1st Mass. Art.

McCrea, Lt. Tully, 1st U.S. Art., Battery I.

McDougall, Col. Archibald L., 1st Brig., 1st Div., 12th Corps.

McFadden, Capt. William, 59th N.Y. Inf.

McFarland, Lt. Col. George F., 151st Penn. Inf.

McFarlane, Lt. Col. Robert, 148th Penn. Inf.

McGilvery, Lt. Col. Freeman, 1st Volunteer Brig., Art. Reserve.

McGroarty, Col. Stephen J., 61st Oh. Inf.

McHenry County, Illinois
Ill. Cav., 8th.

McIntosh, Col. John B., 1st Brig., 2nd Div., Cavalry Corps.

McKean County, Pennsylvania
Penn Inf., 13th Penn. Res., 150th.

McKee, Capt. Samuel A., Jr., 2nd U.S. Inf.

McKeen, Col. Henry B., 148th Penn. Inf.

McMichael, Lt. Col. Richards, 53rd Penn. Inf.

Meade, Maj. Gen. George G., Army of the Potomac Commander.

Medina County, Ohio
Oh. Inf., 8th, 29th.

Mercer County, New Jersey
N.J. Inf., 1st, 6th, 11th.
N.J. Cav., 1st.

Mercer County, Pennsylvania
Penn. Inf., 10th Penn. Res., 57th, 139th, 140th, 142nd, 145th.

Meredith, Brig. Gen. Solomon, 1st Brig., 1st Div., 1st Corps.

Merriam, Lt. Col. Waldo, 16th Mass. Inf.

Merrill, Lt. Col. Charles B., 17th Me. Inf.

Merrimack County, New Hampshire
N.H. Inf., 2nd, 5th, 12th.

Merritt, Brig. Gen. Wesley, Reserve Brig., 1st Div., Cavalry Corps.

Merwin, Lt. Col. Henry C., 27th Conn. Inf.

Messick, Capt. Nathan S., 1st Minn. Inf.

Miami County, Indiana
  Ind. Inf., 20th.
Middlesex County, Connecticut
  Conn. Inf., 14th, 20th.
Middlesex County, Massachusetts
  Mass. Inf., 2nd, 9th, 11th, 13th, 16th,
    19th, 22nd, 28th, 32nd, 33rd.
  Mass. Art., 9th. N.Y. Inf., 74th.
Middlesex County, New Jersey
  N.J. Inf., 1st, 11th.
  N.J. Cav., 1st.
Mifflin County, Pennsylvania
  Penn. Inf., 46th, 49th, 149th.
  Penn. Cav., 1st.
Miller, Lt. Col. Francis C., 147th N.Y. Inf.
Milton, Lt. Richard S., 9th Mass. Art.
Milwaukee County, Wisconsin
  Wisc. Inf., 5th, 6th, 26th.
Moesch, Lt. Col. Joseph A., 83rd N.Y. Inf.
Moffett, Maj. Samuel A., 94th N.Y. Inf.
Monmouth County, New Jersey
  N.J. Inf., 5th.
Monongalia County, West Virginia
  W.V. Inf., 7th.
Monroe County, Indiana
  Ind. Inf., 14th, 27th.
Monroe County, Michigan
  Mich. Inf., 4th, 7th.
Monroe County, New York
  N.Y. Inf., 33rd, 67th, 78th, 108th, 140th.
  N.Y. Art., 1st (L), 5th.
  N.Y. Cav., 8th.
Monroe County, Ohio
  Oh. Inf., 25th.
  W.V. Inf., 7th.
Monroe County, Pennsylvania
  Penn. Inf., 142nd.
Montgomery County, New York
  N.Y. Inf., 43rd.
  N.Y. Art., 1st (K).
Montgomery County, Pennsylvania
  Penn. Inf., 53rd, 68th, 106th.
  Penn. Cav., 1st.
Montour County, Pennsylvania
  Penn. Inf., 6th Penn. Res., 93rd.
Moody, Lt. Col. William H., 139th Penn.
  Inf.
Moore, Maj. John W., 99th Penn. Inf.
Morgan County, Indiana

Ind. Inf., 27th.
Morgan County, Ohio
  Oh. Inf., 25th.
Morgan, Capt. William C., 3rd Me. Inf.
Moroney, Capt. Richard, 69th N.Y. Inf.
Morris County, New Jersey
  N.J. Inf., 7th, 11th, 15th.
Morris, Col. Orlando H., 66th N.Y. Inf.
Morrow, Col. Henry A., 24th Mich. Inf.
Morse, Maj. Charles F., 2nd Mass. Inf.
Mudge, Lt. Col. Charles R., 2nd Mass. Inf.
Muhlenberg, Lt. Edward D., 12th Corps
  Art.
Mulholland, Maj. St. Clair A., 116th Penn.
  Inf.
Munson, Lt. Col. William D., 13th Vt. Inf.
Muskegon County, Michigan
  Mich. Inf., 3rd.
Musser, Lt. Col. John D., 143rd Penn. Inf.

Nantucket County, Massachusetts
  Mass. Inf., 20th.
Neill, Brig. Gen. Thomas H., 3rd Brig., 2nd
  Div., 6th Corps.
Nelson, Capt. Alanson H., 57th Penn. Inf.
Nelson, Maj. Peter A., 66th N.Y. Inf.
Nevin, Col. David J., 62nd N.Y. Inf.
Nevin, Maj. John I., 93rd Penn. Inf.
New Castle County, Delaware
  Del. Inf., 1st, 2nd.
New Haven County, Connecticut
  Conn. Inf., 5th, 14th, 20th, 27th.
New London County, Connecticut
  Conn. Inf., 5th, 14th.
Newton, Maj. Gen. John, 3rd Div., 6th
  Corps.
New York City
  N.Y. Inf., 8th, 10th, 12th, 29th, 39th, 40th,
    41st, 42nd, 43rd, 45th, 52nd, 54th,
    57th, 58th, 59th, 61st, 62nd, 63rd,
    65th, 66th, 68th, 69th, 70th, 71st,
    72nd, 73rd, 74th, 78th, 82nd, 83rd,
    88th, 95th, 102nd, 119th, 145th.
  N.Y. Art., 1st (B), 1st (G), Independent
    Batteries 3rd, 4th, 5th, 6th, 13th, 15th.
  N.Y. Cav., 5th, 6th.
  Penn. Inf., 71st.
  U.S. Regular Inf., 2nd, 3rd, 4th, 6th, 7th,
    12th, 14th.

U.S. Regular Art., 1st (E & G), 1st (H), 1st (I), 1st (K), 2nd (A), 2nd (B & L), 2nd (D), 2nd (G), 2nd (M), 3rd (F & K), 4th (C), 4th (E), 4th (F), 4th (K), 5th (D), 5th (F), 5th (I).
U.S. Regular Cav., 1st, 2nd, 5th.
Niagara County, New York
N.Y. Inf., 49th, 78th.
N.Y. Art., 1st (M).
N.Y. Cav., 8th.
Nichols, Col. William T., 14th Vt. Inf.
Noble County, Ohio
Oh. Inf., 25th.
Norfolk County, Massachusetts
Mass. Inf., 1st, 2nd, 7th, 9th, 12th, 13th, 18th, 20th, 22nd.
Northup, Maj. Charles B., 97th N.Y. Inf.
Northampton County, Pennsylvania
Penn. Inf., 12th Penn. Res., 153rd.
Penn. Cav., 2nd.
U.S. Regular Art., 5th (C).
Northumberland County, Pennsylvania
Penn. Inf., 5th Penn. Res., 46th, 53rd.
Norton, Lt. George W., 1st Oh. Art., Battery H.

Oakland County, Michigan
Mich. Inf., 5th, 7th.
Mich. Cav., 1st, 5th.
Ohio County, Indiana
Ind. Inf., 7th.
Ohio County, West Virginia
W.V. Cav., 1st.
U.S. Regular Cav., 1st.
O'Kane, Col. Dennis, 69th Penn. Inf.
Oliver, Capt. Moses W., 145th Penn. Inf.
Oneida County, New York
N.Y. Inf., 44th, 57th, 78th, 97th, 146th.
Onondaga County, New York
N.Y. Inf., 40th, 122nd, 149th.
N.Y. Art., 1st (B).
N.Y. Cav., 10th.
U.S. Regular Inf., 14th.
Ontario County, New York
N.Y. Inf., 126th.
Ontonagon County, Michigan
Mich. Inf., 16th.
Orange County, New York
N.Y. Inf., 70th, 124th.

Orange County, Vermont
Vt. Inf., 2nd, 4th, 15th.
Orleans County, Vermont
Vt. Inf., 3rd, 15th.
Vt. Cav., 1st.
O'Rorke, Col. Patrick H., 140th N.Y. Inf.
Osborn, Maj. Thomas W., 11th Corps Art. Brig.
Oswego County, New York
N.Y. Inf., 147th.
N.Y. Art., 1st (G),
Otis, Capt. George H., 2nd Wisc. Inf.
Otsego County, New York
N.Y. Inf., 43rd, 76th, 121st.
Ottawa County, Michigan
Mich. Inf., 3rd.
Otto, Lt. Col. August, 58th N.Y. Inf.
Overmyer, Capt. John B., 11th Penn. Inf.
Owen County, Indiana
Ind. Inf., 14th, 19th.
Owens, Capt. Walter L., 151st Penn. Inf.
Oxford County, Maine
Me. Inf., 5th, 16th, 17th.

Packer, Col. Warren W., 5th Conn. Inf.
Pardee, Lt. Col. Ario, Jr., 147th Penn. Inf.
Parke County, Indiana
Ind. Inf., 14th.
Parsons, Lt. Augustin N., 1st N.J. Art., Battery A.
Parsons, Lt. Col. Joseph B., 10th Mass. Inf.
Passaic County, New Jersey
N.J. Inf., 2nd, 5th, 7th, 11th, 13th.
N.Y. Inf., 70th.
Patrick, Col. John H., 5th Oh. Inf.
Patrick, Brig. Gen. Marsena R., General Headquarters.
Paul, Brig. Gen. Gabriel R., 1st Brig., 2nd Div., 1st Corps.
Penobscot County, Maine
Me. Inf., 6th, 7th, 16th, 20th.
Me. Cav., 1st.
U.S. Regular Inf., 17th.
Pennington, Lt. Alexander C. M., Jr., 2nd U.S. Art., Battery M.
Penrose, Col. William H., 15th N.J. Inf.
Perrin, Lt. William S., 1st R.I. Art., Battery B.
Perry County, Pennsylvania

Penn. Inf., 13th Penn. Res.
Philadelphia, Pennsylvania
    Penn. Inf., 23rd, 26th, 27th, 28th, 29th,
      2nd Penn. Res., 12th Penn. Res., 56th,
      61st, 68th, 69th, 71st, 72nd, 73rd,
      74th, 75th, 81st, 82nd, 88th, 90th, 91st,
      95th, 98th, 99th, 106th, 109th, 110th,
      114th, 115th, 116th, 118th, 119th,
      121st, 147th, 150th.
    Penn., Art., 1st (F & G), Independent E.
    Penn. Cav., 2nd, 3rd, 6th, 16th, 18th.
    Del. Inf., 2nd.
    N.J. Inf., 3rd.
    N.Y. Inf., 41st, 71st.
    U.S. Regular Art., 5th (F).
    U.S. Regular Cav., 1st.
Phillips, Capt. Charles A., 5th Mass. Art.
Pickaway County, Ohio
    Oh. Inf., 61st, 73rd.
Pierce County, Wisconsin
    Wisc. Inf., 6th.
Pierce, Col. Byron R., 3rd Mich. Inf.
Pierce, Lt. Col. Edwin S., 3rd Mich. Inf.
Pierce, Col. Francis E., 108th N.Y. Inf.
Pike County, Ohio
    Oh. Inf., 73rd.
Pike County, Pennsylvania
    Penn. Inf., 151st.
Piscataquis County, Maine
    Me. Inf., 6th, 20th.
Pleasonton, Maj. Gen. Alfred, Cavalry
    Corps.
Plumer, Capt. William, 1st Co., Mass. S.S.
Plymouth County, Massachusetts
    Mass. Inf., 7th, 12th, 18th, 32nd.
Portage County, Ohio
    Oh. Inf., 7th.
Porter County, Indiana
    Ind. Inf., 20th.
Potter, Col. Henry L., 71st N.Y. Inf.
Potter County, Pennsylvania
    Penn. Inf., 46th, 53rd, 149th.
Powell, Col. Eugene, 66th Oh. Inf.
Preble County, Ohio
    Oh. Inf., 75th.
Prescott, Col. George L., 32nd Mass. Inf.
Preston County, West Virginia
    W.V. Inf., 7th.
    Md. Inf., 3rd.

Preston, Lt. Col. Addison W., 1st Vt. Cav.
Prey, Col. Gilbert G., 104th N.Y. Inf.
Price, Col. Edward L., 145th N.Y. Inf.
Price, Col. Richard B., 2nd Penn. Cav.
Proctor, Col. Redfield, 15th Vt. Inf.
Providence County, Rhode Island
    R.I. Inf., 2nd.
    R.I. Art., 1st (A), 1st (B), 1st (C), 1st (E),
      1st (G).
    N.Y. Inf., 65th.
    U.S. Regular Inf., 14th.
Pulford, Lt. Col. John, 5th Mich. Inf.
Putnam County, Indiana
    Ind. Inf., 27th.
Pye, Maj. Edward, 95th N.Y. Inf.

Queens County, New York
    N.Y. Inf., 119th.

Racine County, Wisconsin
    Wisc. Inf., 2nd.
Ramsey County, Minnesota
    Minn. Inf., 1st.
Ramsey, Col. John, 8th N.J. Inf.
Randall, Lt. Col. Charles B., 149th N.Y. Inf.
Randall, Col. Francis V., 13th Vt. Inf.
Randol, Capt. Alanson M., 1st U.S. Art.,
    Battery E & G.
Randolph County, Indiana
    Ind. Inf., 19th.
Randolph, Capt. George E., 3rd Corps Art.
Rank, Capt. William D., 3rd Penn. Heavy
    Art., Battery H.
Ransom, Capt. Dunbar R., 1st Regular
    Brig., Art. Reserve.
Rensselaer County, New York
    N.Y. Inf., 104th, 125th.
Revere, Col. Paul J., 20th Mass. Inf.
Reynolds, Capt. Gilbert H., 1st N.Y. Art.,
    Battery L & E.
Reynolds, Maj. Gen. John F., 1 st Corps.
Reynolds, Capt. John W., 145th Penn. Inf.
Rice County, Minnesota
    Minn. Inf., 1st.
Rice, Col. James C., 44th N.Y. Inf.
Richiand County, Ohio
    N.Y. Inf., 59th.
Richland County, Wisconsin
    Wisc. Inf., 5th.

Richmond, Col. Nathaniel P., 1st W.V. Cav.

Rickards, Col. William, Jr., 29th Penn. Inf.

Ricketts, Capt. Robert B., 1st Penn. Art., Battery F & G.

Rigby, Capt. James H ., 1st Md. Art., Battery A.

Rittenhouse, Lt. Benjamin F., 5th U.S. Art. Battery D.

Roath, Capt. Emanuel D., 107th Penn. Inf.

Roberts, Col. Richard P., 140th Penn. Inf.

Robertson, Capt. James M., Reserve Art., Cavalry Corps.

Robinson, Col. James S., 82nd Oh. Inf.

Robinson, Brig. Gen. John C., 2nd Div., 1st Corps.

Robinson, Col. William W., 7th Wisc. Inf.

Robison, Lt. Col. John K., 16th Penn. Cav.

Rock County, Wisconsin
Wisc. Inf., 2nd, 6th.

Rockingham, New Hampshire
N.H. Inf., 2nd.

Rockland County, New York
N.Y. Inf., 95th.

Rodenbough, Theophilus F., 2nd U.S. Cav.

Rodgers, Maj. Isaac, 110th Penn. Inf.

Rogers, Col. Horatio, Jr., 2nd R.l. Inf.

Rogers, Lt. Col. James C., 123rd N.Y. Inf.

Rogers, Lt. Robert E., 1st N .Y. Art., Battery B.

Root, Col. Adrian R., 94th N.Y. Inf.

Rorty, Capt. James M., 1st N.Y. Art., Battery B.

Ross County, Ohio
Oh. Inf., 73rd.

Rowley, Brig. Gen. Thomas A., 3rd Div., 1st Corps.

Ruger, Brig. Gen. Thomas H., 3rd Brig., 1st Div., 12th Corps.

Rugg, 2 Lt. Sylvanus T., 4th U.S. Art., Battery F.

Russell, Brig. Gen. David A., 3rd Brig., 1st Div., 6th Corps.

Rutland County, Vermont
Vt. Inf., 2nd, 5th, 14th.
Vt. Cav.,1st.

Ryder, Capt. Henry W., 12th N.Y. Inf.

Sackett, Col. William, 9th N.Y. Cav.

Sagadahoc County, Maine

Me. Inf., 3rd, 19th.

Saginaw County, Michigan
Mich. Inf., 5th, 16th.
Mich. Cav., 7th.

Salem County, New Jersey
N.J. Inf., 4th, 5th,12th.

Salomon, Lt. Col. Edward S., 82nd Ill. Inf.

Sandusky County, Ohio
Oh. Inf., 8th, 25th, 55th.

Saratoga County, New York
N.Y. Inf., 77th.

Sauk County, Wisconsin
Wisc. Inf., 6th.

Sawyer, Lt. Col. Franklin, 8th Oh. Inf.

Schenectady County, New York
N.Y. Inf., 134th.

Scherrer, Capt. William, 52nd N.Y. Inf.

Schimmelfennig, Brig. Gen. Alexander, 1st Brig., 3rd Div., 11th Corps.

Schoharie County, New York
N.Y. Inf., 95th, 102nd, 134th.

Schoonover, Lt. John, 11th N.J.

Schurz, Maj. Gen. Carl, 3rd Div., 11th Corps.

Schuylkill County, Pennsylvania
Penn. Inf., 96th.
Penn. Art., 1 (F & G).
Penn Cav., 3rd, 17th.
U.S. Regular Art., 5th (K).

Scioto County, Ohio
Oh. Art., 1st (L).

Seaver, Col. Thomas O., 3rd Vt. Inf.

Sedgwick, Maj. Gen. John, 6th Corps.

Seeley, Capt. Aaron P., 111th N.Y. Inf.

Seeley, Lt. Francis W., 4th U.S. Art., Battery K

Selfridge, Col. James L., 46th Penn. Inf.

Sellers, Maj. Alfred J., 90th Penn. Inf.

Seneca County, Ohio
Oh. Inf., 8th, 55th.
N.Y. Inf., 65th.

Seneca County, New York
N.Y. Inf., 126th.

Sewell, Col. William J., 5th N.J. Inf.

Shaler, Brig. Gen. Alexander, 1st Brig., 3rd Div., 6th Corps.

Sharra, Capt. Abram, 1st Ind. Cav.

Sheldon, Lt. Albert S., 1st N.Y. Art., Battery B.

Sherrill, Col. Eliakim, 126th N.Y. Inf.
Sherwin, Lt. Col. Thomas, Jr., 22nd Mass.
   Inf.
Shiawassee County, Michigan
   Mich. Inf., 5th.
   Mich. Cav., 6th.
Sickles, Maj. Gen. Daniel E., 3rd Corps.
Sides, Col. Peter, 57th Penn. Inf.
Sinex, Lt. Col. Joseph H., 91st Penn. Inf.
Sleeper, Capt. Samuel T., 11th N.J. Inf.
Slocum, Maj. Gen. Henry W., 12th Corps.
Smith, Lt. Col. Charles H., 1st Me. Cav.
Smith, Lt. Col. George F., 61st Penn. Inf.
Smith, Capt. James E., 4th N.Y. Art.
Smith, Lt. James J., 69th N.Y. Inf.
Smith, Col. Orland, 2nd Brig., 2nd Div.,
   11th Corps.
Smith, Col. Richard P., Jr., 71st Penn. Inf.
Smith, Lt. William, 1st Del. Inf.
Smyth, Col. Thomas A., 2nd Brig., 3rd Div.,
   2nd Corps.
Snodgrass, Lt. Col. James M., 9th Penn. Res.
Snyder County, Pennsylvania
   Penn. Inf., 6th Penn. Res.
Somerset County, Maine
   Me. Inf., 3rd, 16th, 19th.
   Me. Cav., 1st.
Somerset County, Maryland
   Md. Inf., 1st Eastern Shore.
Somerset County, New Jersey
   N.J. Inf., 3rd, 15th.
Somerset County, Pennsylvania
   Penn. Inf., 10th Penn. Res., 142nd.
Stanford, Capt. Samuel N., 1st Oh. Cav.
Stannard, Brig. Gen. George J., 3rd Brig.,
   3rd Div., 1st Corps.
Stark County, Ohio
   Oh. Inf., 4th, 107th.
St. Clair County, Illinois
   Ill. Inf., 82nd.
St. Clair County, Michigan
   Mich. Inf., 5th.
Steele, Lt. Col. Amos, Jr., 7th Mich. Inf.
Stegman, Capt. Lewis R., 102nd N.Y. Inf.
Sterling, Capt. John W., 2nd Conn. Art,
Steuben County, New York
   N.Y. Inf., 78th, 86th, 107th.
   N.Y. Cav., 6th.
Stevens, Capt. Greenlief T., 5th Me. Art.

Stevens, Capt. Wilber F., 29th Oh. Inf.
Stewart, Lt. James, 4th U.S. Art., Battery B.
St. Lawrence County, New York
   N.Y. Inf., 60th.
   N.Y. Art., 1st (D).
   N.Y. Cav., 6th, 9th.
   U.S. Regular Inf., 11th.
St. Louis County, Missouri
   U.S. Regular Inf., 2nd.
   U.S. Regular Art., 4th (E), 4th (G).
Stone, Col. Roy, 2nd Brig., 3rd Div., 1st
   Corps.
Stoughton, Col. Charles B., 4th Vt. Inf.
Stoughton, Maj. Homer R., 2nd U.S.S.S.
Strafford County, New Hampshire
   N.H. Inf., 2nd.
Stroh, Lt. Col. Amos, 81st Penn. Inf.
Sudsburg, Col. Joseph M., 3rd Md. Inf.
Suffolk County, Massachusetts
   Mass. Inf. 1st, 2nd, 9th, 11th, 12th, 13th,
      19th, 20th, 22nd, 28th, 32nd.
   Mass. Art., 1st, 3rd, 5th.
   Mass. Cav., 1st.
   N.Y. Inf., 70th.
   U.S. Regular Inf., 2nd, 3rd, 6th, 7th, 11th.
   U.S. Regular Art., 1st (E&G), 1st (H), 1st
      (I), 2nd (G), 4th (K), 5th (D).
   U.S. Regular Cav., 5th.
Sullivan County, New Hampshire
   N.H. Inf., 5th.
Summit County, Ohio
   Oh. Inf., 29th.
Susquehanna County, Pennsylvania
   Penn. Inf., 6th Penn. Res., 56th, 141st,
      143rd, 151st.
   Penn. Cav., 17th.
Sussex County, Delaware
   Del. Inf., 1st.
Sussex County, New Jersey
   N.J. Inf., 2nd, 3rd, 7th, 15th.
   N.J. Cav., 1st.
Sweitzer, Col. Jacob B., 2nd Brig., 1st. Div.,
   5th Corps.
Switzerland County, Indiana
   Ind. Cav., 3rd.
Sykes, Maj. Gen. George, 5th Corps.

Taft, Capt. Elijah D., 5th N.Y. Art.
Talbot County, Maryland

Md. Inf., 1st Eastern Shore, 3rd.

Talley, Col. William C., 1st Penn. Res.

Tanner, Capt. Adolphus H., 123rd N.Y. Inf.

Tappen, Maj. John R., 120th N.Y. Inf.

Taylor, Col. Charles F., 13th Penn. Res.

Taylor, Col. John P., 1st Penn. Cav.

Taylor, Lt. Col. William C.L., 20th Ind. Inf.

Thoman, Lt. Col. Max A., 59th N.Y. Inf.

Thomas, Lt. Evan, 4th U.S. Art., Battery C.

Thompson, Capt. James, Penn. Art., Battery C & F.

Thomson, Lt. Col. David, 82nd Oh. Inf.

Thomson, Lt. Col. James M., 107th Penn. Inf.

Throop, Lt. Col. William A., 1st Mich. Inf.

Tilden, Col. Charles W., 16th Me. Inf.

Tilton, Col. William S., 1st Brig., 1st Div., 5th Corps.

Tioga County, New York
N.Y. Inf., 137th.
N.Y. Cav., 5th.

Tioga County, Pennsylvania
Penn. Inf., 6th Penn. Res., 13th Penn. Res., 57th, 149th.
Penn. Cav., 2nd.

Tippecanoe County, Indiana
Ind. Inf., 20th.

Tippin, Col. Andrew H., 68th Penn. Inf.

Titus, Col. Silas, 122nd N.Y. Inf.

Tolland County, Connecticut
Conn. Ind., 14th.

Tompkins County, New York
N.Y. Inf., 64th, 137th.

Tompkins, Col. Charles H., 6th Corps Art.

Torbert, Brig. Gen. Alfred T.A., 1st Brig., 1st Div., 6th Corps.

Touhy, Capt. Thomas, 63rd N.Y. Inf.

Town, Col. Charles H., 1st Mich. Cav.

Trepp, Lt. Col. Casper, 1st U.S.S.S.

Tripp, Lt. Col. Porter D., 11th Mass. Inf.

Trumbull County, Ohio
Oh Inf., 7th.

Turnbull, Lt. John G., 3rd U.S. Art., Battery F & K.

Tuscola County, Michigan
Mich. Inf., 7th.

Tyler County, West Virginia
W.V. Inf., 7th.

Tyler, Brig. Gen. Robert O., Art. Reserve

Ulster County, New York
N.Y. Inf., 71st, 80th, 102nd, 120th.

Underwood, Col. Adin B., 33rd Mass. Inf.

Union County, New Jersey
N.J. Inf., 1st, 2nd, 3rd, 4th, 11th.
N.Y. Art., 6th.

Union County, Ohio
Oh. Inf., 66th, 82nd.

Union County, Pennsylvania
Penn. Inf., 5th Penn. Res., 142nd, 150th.

Upton, Col. Emory, 121st N.Y. Int.

Van Buren County, Michigan
N.Y. Inf., 70th.

Vanderburgh County, Indiana
Ind. Inf., 14th.

Veazey, Col. Wheelock G., 16th Vt. Inf.

Venango County, Pennsylvania
Penn. Inf., 10th Penn. Res., 121st, 142nd.
Penn. Cav., 4th, 16th.

Venuti, Maj. Edward, 52nd N.Y. Inf.

Vermillion County, Indiana
Ind. Inf., 14th.

Vernon County, Wisconsin
Wisc. Inf., 6th.

Vigo County, Indiana
Ind. Inf., 14th.
Ind. Cav., 1st.

Vincent, Col. Strong, 3rd Brig., 1st Div., 5th Corps.

Vinton County, Ohio
Oh. Inf., 75th.

Von Amsberg, George K.H.W., 45th N.Y. Inf.

Von Bourry D'Ivernois, Col. Gotthilf, 68th N.Y. Inf.

Von Brandis, 2 Lt. Hans, 29th N.Y. Inf.

Von Einsiedel, Lt. Col. Heinrich D., 41st N.Y. Inf.

Von Gilsa, Col. Leopold, 1st Brig., 1st Div., 11th Corps.

Von Hammerstein, Lt. Col. Herbert, 78th N.Y. Inf.

Von Hartung, Col. Adolph, 74th Penn. Inf.

Von Mitzel, Lt. Col. Alexander T., 74th Penn. Inf.

Von Steinwehr, Brig. Gen. Adolph, 2nd Div., 11th Corps.

Wabasha County, Minnesota
Minn. Inf., 1st.
Wadsworth, Brig. Gen. James S., 1st Div.,
1st Corps.
Wainwright, Col. Charles S., 1st Corps Art.
Brig.
Walbridge, Col. James H., 2nd Vt. Inf.
Walcott, Lt. Aaron F., 3rd Mass. Art.
Waldo County, Maine
Me. Inf., 4th, 19th.
Walker, Col. Elijah, 4th Me. Inf.
Walker, Lt. Col. Thomas M., 111th Penn.
Inf.
Wallace, Col. James, 1st Md. East. Shore
Inf.
Ward, Col. George H., 15th Mass. Inf.
Ward, Brig. Gen. J. H. Hobart, 2nd Brig.,
1st Div., 3rd Corps.
Warner, Col. Adoniram J., 10th Penn. Res.
Warren County, New Jersey
N.J. Inf., 1st, 7th, 15th.
Warren County, Ohio
Oh. Inf., 75th.
Warren County, Pennsylvania
Penn. Inf., 10th Penn., Res., 13th Penn.
Res., 111th, 145th, 151st.
N.Y. Inf., 74th.
Washington, D.C.
Md. Cav., 1st.
Penn. Cav., 3rd.
U.S. Regular Inf., 4th.
U.S. Regular Art., 1st (K).
U.S. Regular Cav., 2nd, 5th.
Washington County, Maine
Me. Inf., 6th.
Washington County, Maryland
Md. Inf., 1st Pot. Home Brig., 3rd.
Md. Cav., 1st.
Washington County, Minnesota
Minn. Inf., 1st.
Washington County, New York
N.Y. Inf., 123rd.
Washington County, Ohio
Oh. Art., 1st (H), 1st (K).
W.V. Art., 1st (C).
Washington County, Pennsylvania
Penn. Inf., 10th Penn. Res., 140th.
Penn. Cav., 1st, 16th, 18th.
W.V. Cav., 1st.

Washington County, Rhode Island
R.I. Inf., 2nd.
Washington County, Vermont
Vt. Inf., 2nd, 4th, 6th, 13th.
Vt. Cav., 1st.
Washington County, Wisconsin
Wisc. Inf., 26th.
Washtenaw County, Michigan
Mich. Inf., 1st, 4th.
Waterman, Capt. Richard, 1st R.I. Art.,
Battery C.
Watson, Lt. Malbone F., 5th U.S. Art., Bat-
tery I.
Waushara County, Wisconsin
Wisc. Inf., 7th.
Wayne County Indiana
Ind. Inf., 19th.
Wayne County Michigan
Mich. Inf., 1st, 5th, 16th, 24th.
Mich. Cav., 1st, 5th.
U.S. Regular Inf., 17th.
Wayne County, New York
N.Y. Inf., 111th.
Wayne County, Ohio
Oh. Inf., 4th, 107th.
Wayne County, Pennsylvania
Penn. Inf., 6th Penn. Res., 141st.
Penn. Cav., 17th.
Wayne County, West Virginia
W.V. Cav., 1st.
Webb, Brig. Gen. Alexander S., 2nd Brig.,
2nd Div., 2nd Corps.
Weed, Brig. Gen. Stephen H., 3rd Brig., 2nd
Div., 5th Corps.
Weir, Lt. Gulian V., 5th U.S. Art., Battery
C.
Welch, Lt. Col. Norval E., 16th Mich. Inf.
Westbrook, Lt. Col. Cornelius D., 120th
N.Y. Inf.
Westchester County, New York
N.Y. Inf., 49th, 95th.
Westmoreland County, Pennsylvania
Penn. Inf., 11th, 28th, 11th Penn. Res.,
12th Penn. Res., 53rd, 105th, 142nd.
Penn. Cav., 4th.
Wheaton, Brig. Gen. Frank, 3rd Brig., 3rd
Div., 6th Corps.
Wheeler, Col. John, 20th Ind. Inf.
Wheeler, Lt. William, 13th N.Y. Art.

Wheelock, Col. Charles, 97th N.Y. Inf.

White County, Indiana
Ind. Inf., 20th.

White, Lt. Israel, 25th Oh. Inf.

Whiteside County, Illinois
Ill. Cav., 8th.

Whiteside, Capt. Henry, 88th Penn. Inf.

Whittier, Lt. Edward N., 5th Me. Art., Battery E.

Wiebecke, Lt. Col. Charles, 2nd N.J. Inf.

Wiedrich, Capt. Michael, 1st N.Y. Art., Battery I.

Wilkeson, Lt. Bayard, 4th U.S. Art., Battery G.

Willard, Col. George L., 3rd Brig., 3rd Div., 2nd Corps.

Williams, Brig. Gen. Alpheus S., 1st Div., 12th Corps.

Williams, Lt. Col. Jeremiah, 25th Oh. Inf.

Williams, Col. Samuel J., 19th Ind. Inf.

Williston, Lt. Edward B., 2nd U.S. Art., Battery D.

Wilson, Lt. Col. John, 43rd N.Y. Inf.

Windham County, Connecticut
Conn. Inf., 5th.

Windham County, Vermont
Vt. Inf., 2nd, 4th, 16th.
Vt. Cav., 1st.

Windsor County, Vermont
Vt. Inf., 2nd, 3rd, 4th, 6th, 15th, 16th.
Vt. Cav., 1st.

Winegar, Lt. Charles E., 1st N.Y. Art., Battery M.

Winnebago County, Illinois
Ill. Cav., 8th.

Winnebago County, Wisconsin
Wisc. Inf., 2nd, 3rd.

Winona County, Minnesota
Minn. Inf., 1st.

Winslow, Capt. George B., 1st N.Y. Art., Battery D.

Wirt County, West Virginia
W.V. Cav., 1st.

Wister, Col. Langhorne, 150th Penn. Inf.

Wood County, West Virginia
W.V. Cav., 1st.

Wood, Col. James, Jr., 136th N.Y. Inf.

Woodruff, Lt. George A., 1st U.S. Art., Battery I.

Woodward, Lt. Col. George A., 2nd Penn. Res.

Woodward, Capt. Orpheus S., 83rd Penn. Inf.

Woolsey, Capt. Henry H., 5th N.J. Inf.

Wooster, Lt. Col. William B., 20th Conn. Inf.

Worcester County, Massachusetts
Mass. Inf., 2nd, 9th, 13th, 15th, 28th.

Wright, Brig. Gen. Horatio G., 1st Div., 6th Corp.

Wyandot County, Ohio
Oh. Inf., 55th.

Wyoming County, New York
N.Y. Inf., 78th, 136th.
N.Y. Cav., 9th.

Wyoming County, Pennsylvania
Penn. Inf., 12th Penn. Res.

Yates County, New York
N.Y. Inf., 126th.

York County, Maine
Me. Inf., 5th, 10th, 17th.
Me. Cav., 1st.

York County, Pennsylvania
Penn. Inf., 12th Penn. Res., 107th.

Zook, Brig. Gen. Samuel K., 3rd Brig., 1st Div., 2nd Corps.

About the author...

Mr. Raus was born in Cortland, New York on November 10, 1945. He graduated from the State University of New York at Cortland and served in the U.S. Army, 1968-1970. His interest in the American Civil War dates back to knowledge of ancestors serving with the 76th and 23rd New York volunteer regiments. He began his National Park Service career at Gettysburg National Military Park in 1974. Since then he has served as park historian at Fredericksburg National Military Park and is currently chief historian at Manassas National Battlefield Park. He has written numerous magazine articles on various Civil War subjects.

THOMAS PUBLICATIONS publishes books about the American Colonial era, the Revolutionary War, the Civil War, and other important topics. For a complete list of titles, please write to:

THOMAS PUBLICATIONS
P.O. Box 3031
Gettysburg, PA 17325